Concrete reinforcement corrosion

ICE design and practice guides

One of the major aims of the Institution of Civil Engineers is to provide its members with opportunities for continuing professional development. One method by which the Institution is achieving this is the production of design and practice guides on topics relevant to the professional activities of its members. The purpose of the guides is to provide an introduction to the main principles and important aspects of the particular subject, and to offer guidance as to appropriate sources of more detailed information.

The Institution has targeted as its principal audience practising civil engineers who are not expert in or familiar with the subject matter. This group includes recently graduated engineers who are undergoing their professional training and more experienced engineers whose work experience has not previously led them into the subject area in any detail. Those professionals who are more familiar with the subject may also find the guides of value as a handy overview or summary of the principal issues.

Where appropriate, the guides will feature checklists to be used as an *aide-mèmoire* on major aspects of the subject and will provide, through references and bibliographies, guidance on authoritative, relevant and up-to-date published documents to which reference should be made for reliable and more detailed guidance.

ICE design and practice guide

Concrete reinforcement corrosion

From assessment to repair decisions

Peter Pullar-Strecker

Published by Thomas Telford Publishing, Thomas Telford Ltd, 1 Heron Quay, London E14 4JD
http://www.thomastelford.com

First published 2002

Distributors for Thomas Telford books are
USA: ASCE Press, 1801 Alexander Bell Drive, Reston, VA 20191-4400, USA
Japan: Maruzen Co. Ltd, Book Department, 3–10 Nihonbashi 2-chome, Chuo-ku, Tokyo 103
Australia: DA Books and Journals, 648 Whitehorse Road, Mitcham 3132, Victoria

A catalogue record for this book is available from the British Library

Classification
Availability: Unrestricted
Content: Recommendations based on current practice
Status: Refereed
User: Practising civil engineers and designers

ISBN: 07277 3182 3

Typeset by Gray Publishing, Tunbridge Wells, Kent
Printed in Great Britain by Bell and Bain, Glasgow

Contents

1. Introduction and scope

Uncontaminated uncarbonated concrete is one of the most effective materials for protecting steel from corrosion because steel is passive in this highly alkaline environment. The earliest users of reinforced concrete (it was patented by Wilkinson in 1854) did not know how lucky they were; they looked on concrete cover as a sort of mackintosh that protected reinforcement by keeping the weather out. To ensure that the steel was properly embedded they specified wet mixes, and, of course, these carbonated quickly. By about 1930 this fashion had passed and since the publication of the first British Code of Practice in 1934 durability has not been a problem in the overwhelming majority of construction. In the small number of cases where durability has turned out to be inadequate, the corrosion of reinforcement has been by far the most common problem. Fascinating accounts of these early days are given in reference [1].

In the 1960s and 1970s the increased use of de-icing salts on highways in the USA and the construction boom in Arabian Gulf area highlighted the problems of reinforcement corrosion caused by chlorides, especially in hot arid regions. This triggered intensive world-wide research into the causes and repair of reinforcement corrosion, which in turn led to a vast output of research papers, conferences and publications on the subject. Available reference material includes original papers recording developments in understanding the fundamental science and mechanisms, papers sharing practical experience of repair applications and site trials, review papers summarizing the present state of knowledge and books and guides which share with the reader their authors' comprehensive knowledge and experience on every aspect of the subject. A recent and important addition is a suite of European standards devoted to the repair and protection of concrete affected by reinforcement corrosion. A very small selection from the available material is included in this guide's Bibliography in the hope that any of these publications will be easy to find in the libraries of the civil and structural engineering institutions and will lead the more interested reader on to other more specialized publications from their own lists of references.

This guide does not attempt even to summarize the vast body of present knowledge. Rather, it concentrates on 'the need to know'. It assumes that 'the need to know' arises from a 'need to do', and focuses on the client's need to make decisions about what to do. The guide has been developed from course notes that the author wrote for a series of continuing professional development courses for engineers in the UK, the Arabian Gulf Region and the Far East and tries to answer the questions that practitioners ask, and perhaps some of those they may wish they had asked.

Although the guide aims to be concise and practical, it tries not to shut out those who are developing a deeper interest in reinforcement corrosion, either with the possibility of becoming specialized practitioners in this fascinating and commercially important area, or with the intention of becoming involved in research and development to add to the store of useful knowledge and the development practicable applications.

However practical their approach, investigators of corrosion problems in reinforcement must have at least a basic understanding of corrosion mechanisms to assess the dangers, and some cases even to see the symptoms: the great landscape painter and master of observation John Constable said 'I see only what I understand' and to some extent this applies to all of us. Without understanding it is impossible to judge whether a crack is a cause of corrosion or its result, or neither, or why repairing one area can cause corrosion in a neighbouring one. In the detective work of investigation and assessment, the good sleuth has to know what must be considered and what can be ignored.

Equally, a working knowledge of how structures are designed and built is essential for corrosion experts assessing where the dangers lie. Knowledge of how and where bars are linked and lapped and fixed, how formwork is fixed and how concrete is poured can lead the shrewd investigator straight to the most likely trouble spots. 'How was this built?' is always an important question to ask when looking for defects.

The scope of this guide includes the basic considerations for the investigation, assessment, protection and repair of reinforced concrete affected by corrosion. It does not include post-tensioned concrete, though the general principles can be applied to reinforcement associated with post-tensioning. Environmental conditions cover natural exposure but not the special hazards sometimes generated by industrial processes.

2. Understanding corrosion

2.1 The production of electricity by electrochemistry

Rusting is an electrochemical process very like the one that produces power in an ordinary torch battery. A useful model for explaining the terms and mechanisms is the practical power-producing **cell** invented by Volta in about 1800. It had one copper plate (**electrode**) and one zinc plate. When the plates were immersed in dilute sulphuric acid (the **electrolyte**), the zinc plate became negative (it was the **anode**) and the copper plate positive (it was the **cathode**), with a potential difference between them of approximately what we now call one volt (see Figure 2.1). When the plates were connected together the cell produced an electric current. Electrons (the negative carriers of the current) left the cell at the anode (this is what *defines* an anode), which slowly dissolved, and returned to the cell via the cathode, which was not attacked. Charged part-molecules (**ions**) in the acid (the electrolyte) completed the circuit.

It can be difficult to identify anodes and cathodes and their polarity. Whether an electrode is acting as an anode or as a cathode depends on whether the cell is *generating* current, as in the case of the Volta cell, or *consuming* current, as in the case remedial techniques like impressed current **cathodic protection.**

In a cell which is *generating* current, the anode is the *negative* electrode because electrons leave the cell there, but where a current is passed through a cell from an

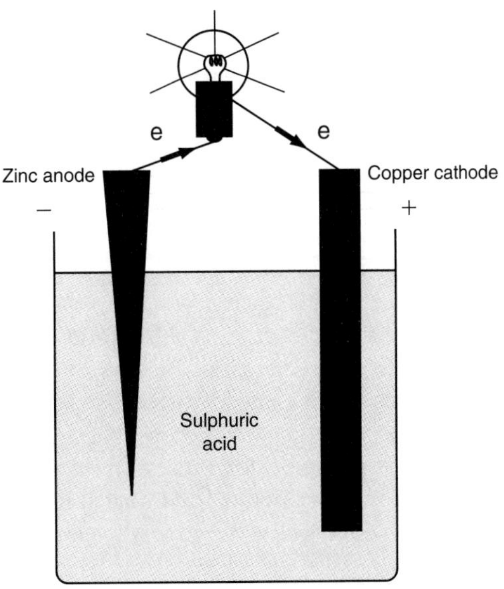

Figure 2.1 Volta's cell: electrons leave the cell at the anode, which then dissolves.

Table 2.1 Standard electrode potentials of some common metals in relation to hydrogen

Gold	+1.68
Mercury	+0.86
Silver	+0.80
Copper	+0.35
Hydrogen	**0.00**
Nickel	−0.22
Iron	−0.44
Zinc	−0.76
Aluminium	−1.66
Magnesium	−2.37
Lithium	−3.30

external source, as in impressed current cathodic protection, the electrode connected to the *positive* side of the power source is the anode in the cell because the external power *draws* the electrons out from the cell here. In both cases, indeed in all cases, the electrode where electrons leave the cell is the anode and it is the anode that is attacked.

Different metals, alloys and compounds produce different potentials when they are placed in an electrolyte. If this is done under standardized conditions, the potentials form a series, known as the **electrochemical series**, which has fixed values.

Each metal is able to act as a cathode in an electrolytic cell with any metal that is below it in the series, and thus cause any of the lower metals to dissolve. Hydrogen is the element used as the reference point for potentials. Lead peroxide is a familiar example of a compound acting as an electrode: it is the positive electrode in the ordinary lead–acid car battery, where lead is the negative electrode (see Table 2.1).

Standard electrode potentials are measured at a fixed temperature and in electrolytes which are solutions of known concentration of the electrode's own ions. At other temperatures, in other electrolytes, or in electrolytes of different concentrations, the potentials will be different. The availability of oxygen also affects the potential of an electrode.

Thus it is quite possible to have a working cell with the same material for both anode and cathode, and with the same electrolyte in contact with both, provided that the concentration, the temperature or the **oxygen** availability are different at each electrode. It is in this way that electrolytic **cells** can occur in reinforced concrete that contains only steel as electrode material and only **pore fluid** as the electrolyte (see Figure 2.2).

The electrical nature of corrosion makes it possible to use electrical measurements and processes for a variety of purposes that range from detecting anodic reinforcement to controlling corrosion. **Electrode potential** measurements on the surface of the concrete are used to locate areas where reinforcement is **active** and in a condition to corrode; an electric current is used in **cathodic protection** to keep the reinforcement at a potential at which it cannot corrode, in **electrochemical chloride extraction** to move chloride ions away from the reinforcement and in **electrochemical realkalization** to generate hydroxyl ions near the reinforcement. In resistive control, the electric current is virtually eliminated by reducing the moisture content of the concrete.

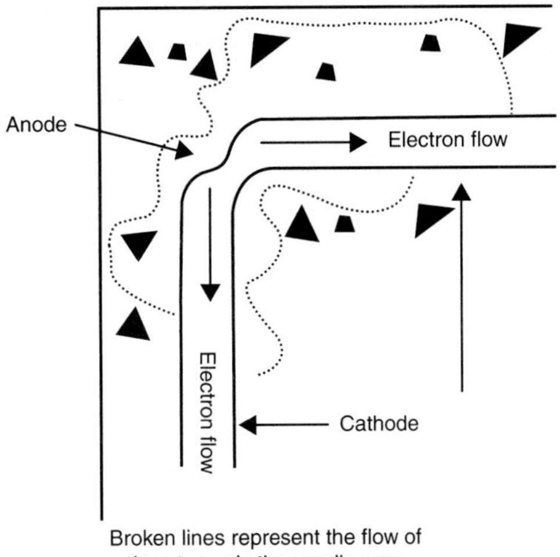

Figure 2.2 A corrosion cell in concrete: electrons leave the reinforcement at the anode, which then dissolves.

2.2 Corrosion of steel in concrete

Under most conditions, corrosion does not take place, or at least not at a rate that has any practical effect on the **service life** of the structure. Five mechanisms can prevent corrosion or control the **corrosion current**, and at least one of them must be present in any repair method which is used to control corrosion. They are **passivity**, **resistive control**, **cathodic control**, **anodic control** and **cathodic protection**, the most important being passivity.

If corrosion is neither absent as a result of passivity nor controlled to a very low level by **polarization** or **resistivity**, corrosion of some kind will take place. Under normal conditions corrosion usually takes the form of **general corrosion** that results in familiar uniform rusting. Here **anodic** and **cathodic** areas are too close together to be seen separately and rust can grow to a volume several times that of the steel from which it was formed. The result is that unless the reinforcement is of an exceptionally small cross-section (steel fibres for example) the growing rust will eventually **crack** the cover and may also cause **delamination**.

Where there is a significant concentration of **chloride contamination** in wet concrete and little oxygen is available at corroding anodic areas, soluble **iron compounds** may escape through cracks or pores in the concrete, leaving the reinforcement deeply **pitted** sometimes without there being any visible sign of damage on the concrete surface (see Figure 2.3). In this case when the concrete is broken open, green or black partially oxidized corrosion products can be seen to have spread through the concrete.

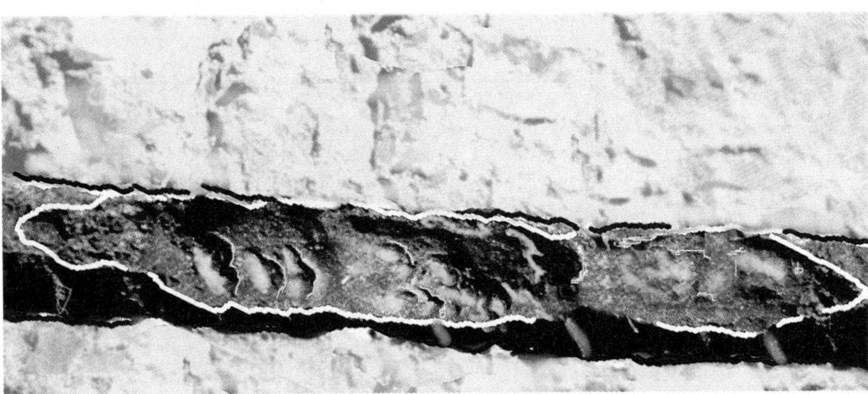

Figure 2.3 This pitting corrosion was not suspected until the wet salty concrete was broken out.

Figure 2.4 Corrosion products in concrete saturated with sea-water soon after it was broken open.

This type of corrosion is often called **black rust**, though it turns rusty-red in a few minutes when it is exposed to air. **Rust staining** that has spread from reinforcement is often an indication that the cause of corrosion is chloride contamination (see Figure 2.4).

2.2.1 Control of corrosion by passivity

Steel which is *completely* surrounded by uncontaminated and uncarbonated cement paste is protected by a strongly **alkaline environment** where the **pH** is 11 or more. In this environment, a thin film of **ferric oxide (gamma Fe_2O_3)** forms on the surface of the steel and prevents the steel from corroding (see Figure 2.5). Steel that is in this condition is said to be **passive**. Passivity will be lost if the alkalinity of the concrete is reduced by **carbonation**, that is by carbon dioxide (from the air) reacting with the main alkaline constituent of cement, **calcium oxide**, and any water in the concrete pores to form calcium carbonate. **Calcium hydroxide**, which is the result of the reaction

Figure 2.5 Corrosion occurred only where there was a break in the cement paste coating.

between calcium oxide ('portlandite') and water, is soluble to little more than 2% and the reserves of calcium oxide which remain undissolved continue to provide protection (buffering) for a considerable time during the carbonation process, especially in rich mixes (see Figures 2.6 and 2.7).

Even if the concrete is uncarbonated and therefore strongly alkaline, passivity can also be lost if **chlorides** are present at a sufficiently high concentration. Hauseman calculated the widely accepted threshold ratio for corrosion as 0.6 of chloride ion to 1.0 of hydroxyl ion, but the later work by Pourbaix, Page and others has shown that this is an over-simplification (see reference [2]). A table of threshold values based on UK experience is given later in Table 4.2. Carbonation reduces the concentration of hydroxyl ions and therefore the concentration of chloride needed to destroy passivity. Chloride ions appear to act as catalysts and their effect continues undiminished by the amount of corrosion product that has been produced. Part of any chloride that is

Figure 2.6 Perfectly protected reinforcement in a 50-year-old second world war Mulberry harbour unit on the beach at Arromanches, France. The concrete is less than 25 mm thick and is said to have had a cement content of 650 kg/m^3.

Figure 2.7 Massive construction lasts: first world war pill box in Belgium.

present in the mix combines with calcium aluminates (mainly C_3A) in the cement but is released when concrete is carbonated or when acid extraction is used in preparing samples for chemical analysis. Sulphate-resisting cements are the types most *likely* to be found low in C_3A, but specifications for other cement types do not in fact include a *lower* limit for C_3A. See later for Table 4.2 that relates chloride concentrations to the UK experience of the likelihood of corrosion.

2.2.2 Resistive control of corrosion rates

If **passivity** is lost, parts of the steel can become **anodic** and dissolve unless other mechanisms prevent corrosion or control it to an insignificant rate. The most usual controlling factor is the electrical 'resistance' of the concrete, termed **resistivity** and defined as the resistance between the opposite faces of a 1 cm cube. (The units of resistivity are ohm-cm, that is dimensionally ohms per centimetre per unit area of cross-section.) **Conductivity** in concrete is due to the presence of (charged) **ions** in the **pore fluid in inter-connected capillaries** in the concrete. The **corrosion current** is restricted by the limited pore fluid in the small number of fine inter-connected capillaries. In very dense or very dry concrete, little current will flow (so corrosion will be slow) even if passivity has been lost. Corrosion is then said to be under **resistive control**. This mechanism is often the main factor controlling corrosion in concrete that is permanently dry, for example inside heated buildings (relative humidity of 60–70% at the reinforcement is often cited as the threshold for corrosion at a significant rate), but even saturated concrete can have a high resistivity if the inter-connected capillaries are few and fine, as they will be in low water : cement ratio concrete, or in concrete made with some types of **additive** like silica fume. In existing concrete, recent research has shown that a useful degree of moisture reduction (reduction in **relative humidity**) can be achieved with coatings and surface treatments that allow water vapour to escape but do not allow water as liquid to enter (see Section 5.10). The resulting increase in resistivity can control corrosion to an acceptable level in uncontaminated carbonated concrete though not in concrete where there is a significant level of chloride contamination. The likelihood of corrosion occurring at a significant rate in concrete has been assessed by many observers in relation to electrical resistivity. A widely accepted classification is given in Table 2.2 [3].

2.2.3 The role of oxygen, and cathodic control of corrosion

The corrosion process cannot take place without oxygen at the cathode though none is needed at the corroding anode. In some situations where the steel is not passive, the corrosion current can be limited by **polarization** of the cathode through lack of oxygen. If oxygen access is very restricted at the part of the reinforcement that would potentially form the cathodic electrode of a cell, current will stop flowing when sufficient hydrogen has been generated at the cathode, just as it did in Volta's cell where the copper plate had to be lifted out of the acid from time to time to keep the cell working. Because current can flow at the anode only when there is an equal flow at the cathode, corrosion will stop even if the potentially anodic area is salt-contaminated

Table 2.2 Risk of corrosion related to resistivity

Resistivity (ohm-cm)	Corrosion risk
<5000	Very high
5000–10 000	High
10 000–20 000	Moderate to low
>20 000	Insignificant

or in a condition to corrode for some other reason. In this case corrosion is said to be under **cathodic control**. This condition can often be seen in isolated units that are completely **saturated**, even with **sea-water**, when there is no electrical connection with units which have oxygen access. Failure to understand this mechanism was one of the reasons why some early researchers of reinforcement corrosion in sea-water were unable to agree whether or not chlorides caused corrosion: many of their specimens were completely submerged. Research for the Concrete-in-the-Oceans programme showed that where the concrete is undamaged a relatively large cathodic area is needed to supply a significant current to a corroding anode because of the restricted oxygen supply at cathodic areas. Research has not yet shown whether a useful degree of cathodic control could be imposed by coating the surfaces of unsaturated concrete to restrict oxygen access.

2.2.4 Coating reinforcement, and the anodic control of corrosion

In theory at least, although not always in practice, corrosion can be also prevented by coating the reinforcement (and therefore any part of it which might become anodic) with an electrically insulating waterproof **barrier coating**, such as fusion-bonded epoxy resin. This is **anodic control**. In aggressive environments the use of coated reinforcement to prevent corrosion has produced disappointing results, probably because if any part of the coating has been damaged, the ideal protective coating, i.e. the intimate coating of uncontaminated alkaline cement paste, is not there to protect the steel.

2.2.5 Imposing cathodic protection

The fifth method of protection differs from the others in that it is in a sense artificial and it is imposed from outside the system. **Cathodic protection** is a method in which all the reinforcement is kept cathodic in relation to an external anode (see Section 5.11). The necessary power is usually taken from the local electricity supply but can be generated galvanically by a **sacrificial** anode that gradually wastes away. 'Cathodic protection' of a different kind sometimes also occurs in **chloride-contaminated** reinforced concrete when strongly corroding anodes prevent corrosion in neighbouring *potentially* anodic areas (sometimes called an **incipient anode**) because they are currently cathodic to these anodes. This is most likely to happen in relatively wet conditions where the low resistivity of the concrete provides an effective return path for the protective current. If one of the strongly corroding anodes is repaired, say by concrete patching, some of the areas that were cathodic can lose their protection, become anodic and themselves start to corrode. Cathodic protection imposed from outside the system can keep all of the active and incipient anodes cathodic in relation to an anode system attached to the surface of the concrete. The practical applications of cathodic protection are covered in Section 5.11.

2.2.6 Using external power to extract or introduce ions

Two other uses of power from outside the system are the extraction of chloride ions from **chloride-contaminated** concrete and the creation of hydroxyl ions (and sometimes also the introduction of alkaline metal ions) in **carbonated concrete**. In these processes temporary anodes are attached to the surface of the concrete and power is applied between them and the reinforcement for a short period until the changes have been effected. The practical applications of **electrochemical chloride extraction** and **electrochemical realkalization** are covered in Sections 5.12 and 5.13.

3. Initial investigation and assessment of concrete structures

3.1 Scope and brief for the initial investigations

Possible triggers for assessing corrosion in a reinforced concrete structure include visible signs of trouble, a proposed change of use or owner, or a routine health check.

Assessment can cover a wide range: simple visual inspection of easily accessible parts by an experienced investigator if the structure appears to have little or no **damage**; limited examination of sample areas to establish the seriousness of any more extensive visible damage; or detailed testing, sampling and exploratory removal of concrete to enable a **repair specification** to be prepared.

To keep costs under control, assessment should normally be done in stages with each stage having a limited scope and purpose which is clearly defined in a **brief** agreed between the **client** and the investigator (see 'Checklist of briefs for investigations'). **Initial investigations** should normally made as simply as possible but where the investigator is working a long way from home, or where access is difficult for other reasons, it can pay to be ready to extend the initial investigation so as to avoid the risk of needing to make an early return visit.

Apart from the desirable but uncommon routine health check, where the client will probably have a standard specification for the work, any assessment will be efficient and **economical** only if the investigator clearly understands *why* the assessment is to be made and *what* the client is hoping to achieve. Without a clear brief, the amount of data an investigator could collect and the range of tests that could be made at any stage of investigation are almost unlimited. With a clear brief, the investigator can limit activity to collecting and reporting only the information necessary to achieve the specified objectives at an economical cost. A clear brief, even it is only a simple letter, also helps to control the investigator's risks.

If the investigation is being carried out to provide **evidence** for use in **arbitration** or **litigation** in addition to, or as an alternative to, providing the information needed to make repair decisions, it is best to know this from the start. Collecting evidence needs a level of meticulous recording, accuracy and detail that goes far beyond what is required even to specify the repair itself. After all, someone is going to be well paid to find weaknesses in the opposition's case.

In this guide, attention will be drawn to items that have special relevance to collecting evidence.

The sole purpose of any stage of assessment is to enable the client to make a **decision** about what to do next. Typically the **options** will be to do nothing; to commission a more **detailed investigation**; to change the use of the structure; to protect, preserve or repair it; to improve it; to rebuild it; to sue, or to sell.

To make a rational decision the client needs to know whether there is a problem that needs action; if there is, what choices there are and what further information will be needed to evaluate the options. The client will also need an estimate of the cost of going to the next stage.

It is important that the investigator is always aware that these decisions must be made by the client, albeit with the help of the investigator's knowledge and experience. It is unreasonable to expect that every client will take a close interest in the technicalities of the investigations, but it is helpful if the client or the client's representative is present for at least part of the initial investigation so that the client has some personal knowledge of the purpose, scope and limitations of the investigation.

Initially the investigator should be able to discover whether any part of the structure is already **defective** as a result of reinforcement corrosion; if it is not, whether any part is likely to become defective during the proposed **service life** of the structure, and whether the cause of any present or potential damage is the common one of **carbonation** or whether it could include the more serious problem of **chloride contamination**. Later stages of the assessment, if any are needed, will look at the types and quantities of damage and the possible repair options.

3.2 Where to look for corrosion damage and how to recognize it

Safety. Before working on a site, if there appears to be any risk to people or property in or around a structure (falling pieces of concrete are the most usual danger: they can and have caused death), the client must be advised and must decide what action to take to avoid the risk of personal injury or damage to property. The advice must be confirmed in writing and (as always) a copy should be filed.

If deterioration caused by corrosion is spotted early enough, preventive measures like **coating** or **surface treatment**, which are much cheaper than repair, can sometimes be used to control or even prevent its development. It is therefore important that the earliest signs of corrosion are identified however difficult they are to spot. It helps to know the **vulnerable locations**, and a background of knowledge and experience of how reinforced concrete structures are designed and built (and how this was done at the date of construction) is invaluable.

Tip. 'How was this built?' is a question always to keep in mind.

Early signs of **deterioration** from reinforcement corrosion are most likely to be seen where reinforcement is especially close to the surface, where compaction has been

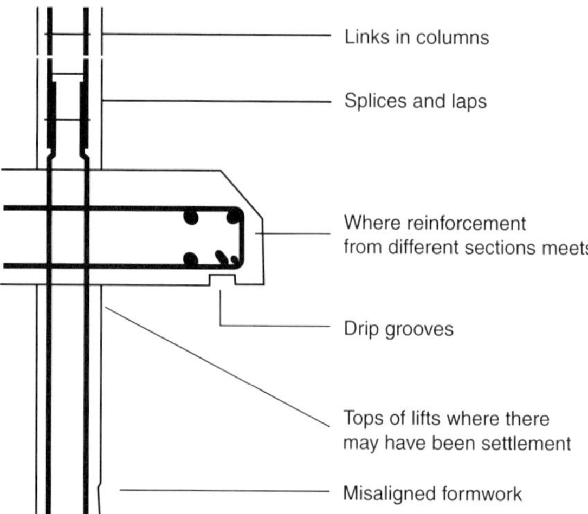

Figure 3.1 Reinforcement that is close to the surface is especially likely to have lost protection.

difficult, where the surface has been subject to frequent wetting and drying or where chlorides have penetrated the concrete. (See the 'Checklist of **vulnerable locations**' given at the end of Chapter 4 and Figure 3.1.)

Equipment for initial surveys usually consists of no more than common tools plus a spray-gun with **phenolphthalein indicator solution**, a simple **half-cell** and a basic **cover meter**. Advice on how to carry out tests is given in Chapter 4. (See the 'Checklist of useful equipment for investigations' given at the end of Chapter 4.)

3.2.1 Cracking caused by carbonation and chloride contamination

Conditions at the corroding area determine the type of corrosion that will occur and what signs will be visible. The most common early sign of corrosion is cracking, though to have some cracking on the surface of reinforced concrete is quite normal. Normal includes transverse shrinkage cracks formed early in the **hardened concrete**, transverse **controlled cracks in tension zones** which can occur at any time, and **plastic settlement cracks** and **plastic shrinkage cracks** formed before the concrete has hardened.

The **transverse cracks** which are normal in reinforced concrete are controlled by the reinforcement and are usually less than 0.3 mm wide. They are caused by shrinkage, **thermal contraction** or **structural loading** but not usually by corrosion. They seldom cause reinforcement corrosion, even in **salty environments** probably because the small area of reinforcement exposed by a controlled transverse crack is usually adequately coated with cement paste and is unlikely to be large enough to support a corrosion cell with the required **oxygen** access to the cathode.

Exceptions occur where transverse cracks coincide with adjacent bars in a *lower* layer running at right angles. This can happen if the lower bars have acted as **crack inducers**. These secondary *longitudinal* cracks can be vulnerable to carbonation or salt penetration along their length, though they have not been *caused* by corrosion. (See Section 5.5.3 for comment on protection and repair.)

Plastic shrinkage cracks form before the concrete has hardened and are caused by rapid loss of water from the surface of the fresh concrete. They affect mainly large surfaces like slabs and the cracks normally run diagonally across the area and are tapered at both ends. These cracks are easy to recognize (see Figure 3.2). They are not

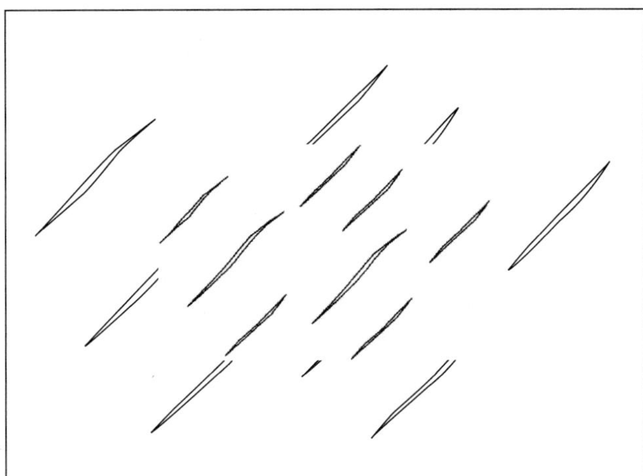

Figure 3.2 Plastic shrinkage cracks are not a sign of corrosion, but they can help to cause it.

caused by corrosion but can be a cause of it. Where the concrete surface is exposed to salt, the cracks can hold water and salt and allow **chlorides** to penetrate close to the reinforcement.

Plastic settlement cracks are not caused by corrosion, but they can cause it because, unlike most other plastic cracks, they are parallel with the reinforcement and furthermore lie over reinforcement which is often not protected by cement paste on the underside because **bleed** water has removed it. Plastic settlement cracks occur as a result of aggregates in over-workable concrete settling in the formwork after the concrete has been compacted, and cracking the concrete where it hangs over the reinforcement as it stiffens (see Figure 3.3). These cracks cannot, of course, be seen in intermediate lifts unless corrosion has already resulted in visible **defects**, but their likely positions can be estimated from the positions of lift lines and they should be seen as potential trouble spots, especially where there is chloride contamination. Sea and harbour walls are quite often affected by corrosion caused by this type of crack.

Carbonation of the cover affects all concrete to some degree and is by far the most common cause of reinforcement corrosion, but contamination with chlorides is a more serious one. Sometimes rapid corrosion is caused by a combination of both causes. If concrete is carbonated down to the reinforcement or contaminated with chlorides in relatively dry conditions, the form of corrosion is usually **uniform rusting (general corrosion)** over a relatively large area of the steel. **Anodic** and **cathodic** points are too close together to be seen as separate areas. In wet concrete, any cracking caused by chloride-induced **pitting corrosion**, is easy to recognize. (See Section 3.2.4 below.)

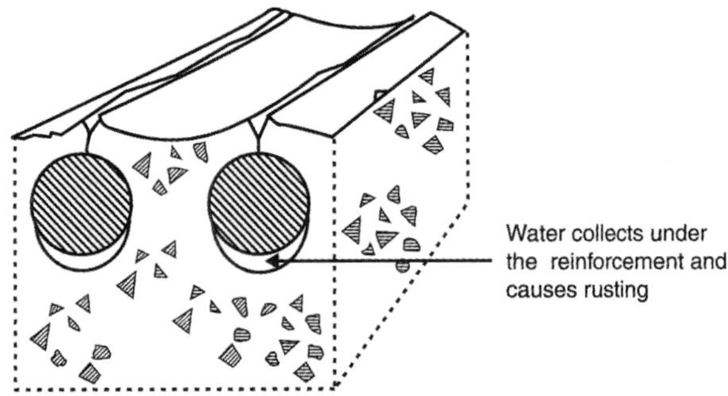

Figure 3.3 Plastic settlement cracks form if very workable concrete settles in the formwork: they can cause corrosion.

In relatively dry concrete, cracking resulting from the two causes looks much the same which makes it difficult to know whether chloride contamination is present in these conditions. The differences in the repair methods that would be appropriate and the cost implications make it essential to find out whether chlorides are present. Knowledge of the history of the structure, its location in relation to the sea or salted highways, and the use of **electrode potential mapping** can provide an indication of whether chloride contamination is likely to be present. Confirmation will need sampling and **chemical analysis** not usually undertaken in initial surveys and these subjects are dealt with in Chapter 4. At the initial investigation stage the investigator will be able to report on the defects that were found with much more confidence than on the absence of any problem (see Section 3.3).

Uniform rusting can be taking place for several years before it eventually produces enough rust to cause cracking and (later) delamination or spalling of the surface concrete. The time taken for this to happen depends on the severity of exposure.

The type of crack which is almost always the first sign of uniform rusting is a fine **longitudinal crack** close to the position of the rusting reinforcement and running more-or-less parallel with it. A **cover meter** is the best instrument for confirming that reinforcement lies under a crack, but the investigator should suspect that reinforcement lies below if cracks are spaced at intervals that correspond to conventional reinforcement spacings. Marking the positions of any visible cracks against a measuring tape (a surveyor's folding rule is more convenient if working single-handed) over a metre or two and then marking the expected positions of any 'missing' cracks often reveals very fine cracks which were not noticed at first: cracks as fine as about 0.05 mm or even less can be seen by someone who knows where to look (see Table 3.1 below). Because cracks hold water, they are easier to see when concrete is drying out; spraying with water and waiting until the surface is just dry can help to reveal them; drying a wet surface with a blow lamp or hot air gun can have the same effect.

It can be useful to mark the ends of cracks permanently on the concrete during an **initial investigation** so that any extension can be identified in subsequent inspections if a decision is taken to postpone repair. This also makes crack lengths easier to estimate from **photographs**. Similarly, **classifying cracks by width** may allow increases in crack width to be identified in later surveys. An optical crack width gauge or a vernier calliper (used as a visual comparator) are sufficiently precise. An

Table 3.1 Classifying cracks by width

Description	Width	Visibility	Significance of transverse cracks
Hair or very fine cracks	0.05–0.1 mm	Noticeable when drying out: once noticed, visible to unaided eye in good light	Unlikely to cause corrosion
Fine cracks	0.1–0.3 mm	Noticeable to unaided eye	Unlikely to cause corrosion except in severe exposure
Medium to wide cracks	0.3–1.0 mm	Easily noticeable; both edges of the crack may be visible	Wider than cracks acceptably controlled by reinforcement; could cause corrosion
Wide to very wide cracks	>1.0 mm	Both edges of the crack easily visible; an estimate of the crack width should be recorded	Corrosion danger increasing with crack width

interesting discussion of the significance of cracking when assessing corrosion is given in reference [4].

> *Tip.* After marking features of interest on the surface of the concrete, add information that identifies the location and then photograph it. It is surprising how much more will be seen from photographs in the comfort of the office, and surprising how quickly one can forget the exact location of an individual photograph. (See Section 4.3.)

A cover meter will allow the thickness of cover to be estimated (even if it is set for an **assumed bar diameter**) and low values (<38 mm) may suggest that cracking is likely to be caused by uniform rusting in carbonated concrete. However, **concrete quality** has the dominating influence on how quickly carbonation will reach the reinforcement and the locally destructive test for **carbonation depth** with **phenolphthalein indicator** is the only way to be sure (the use of both tests is explained in Chapter 4). Except where the phenolphthalein test needs to be used, chipping away cover so that the reinforcement can be seen is not usually a good idea during preliminary visual inspections: the reinforcement will probably not look significantly rusty and the damage to the cover will have to be made good reasonably quickly.

Tapping around the region of the crack with a light hammer (a 4-oz pin hammer is ideal) can sometimes reveal the hollow sound of **delamination** (see below) even at the earliest stage. As the rust grows, cracks become wider and easier to see. Eventually a second roughly parallel crack may form and allow the cover to delaminate or even **spall** from the parent concrete. Once the cover has come off, the reinforcement can usually be seen without breaking out further concrete.

3.2.2 Delamination and spalling

Spalling of the concrete surface is an obvious sign of advanced reinforcement corrosion. Spalling is caused by the growing **rust** first cracking and then pushing off part of the concrete cover (see Figures 3.4 and 3.5). A less obvious sign of corrosion is the **delamination** that precedes spalling and results in hollowness below the surface. Where a series of (usually parallel) reinforcing bars rusts, the concrete can split neatly along the plane of the reinforcement and after a time the rust may grow enough to detach the cover in large sheets.

Before concrete has spalled, delamination cannot be seen but it can be heard. If the concrete surface is tapped gently with a small hammer, the hollow sound of delamination soon becomes easy to recognize except in noisy surroundings. Delamination in large areas can quickly be identified by simply scratching across the concrete surface lightly with a hammer-head or piece of reinforcement. A hollow 'roaring' sound indicates delamination.

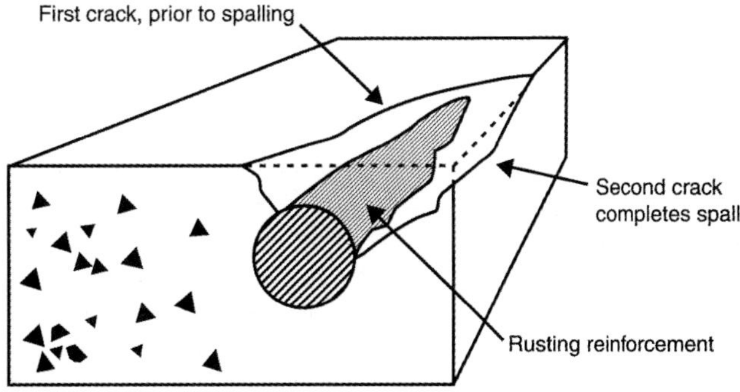

Figure 3.4 How spalling develops over rusting reinforcement.

Figure 3.5 Typical spalling over links with low cover in a British 1960s office block.

> *Warning.* Delamination should be treated with great respect. Hard hammering can bring the cover down in pieces big enough to cause serious injury.

3.2.3 Examination of the reinforcement

Where concrete has **spalled** and the reinforcement can been seen easily, examining the surface of the concrete immediately around the exposed reinforcement can often indicate whether the rusting was caused by **chloride contamination** or by **carbonation.** Rusting caused by carbonation (general **rusting**) seldom stains the concrete very much, whereas rusting caused by chloride contamination, especially in wet conditions, tends to be '**pitting corrosion**' and usually starts as **a soluble corrosion product** (an iron compound in solution) which spreads into the concrete for a short distance, resulting in **rust staining** when it becomes oxidized (see Figure 3.6). Visible pitting of the steel is another indication of chloride-induced rusting.

On an external concrete surface, unsightly rust staining is more often the result of **aggregate contamination** (with iron pyrites, for example), or nails or pieces of tying wire left lying on soffit formwork, but here the cause will be obvious from an inspection

Figure 3.6 Typical staining from corrosion in salt-contaminated concrete. Support leg of a British tunnel roadway.

Figure 3.7 This staining is caused by pyrites inclusions in the aggregate, not by reinforcement corrosion.

of the concrete surface. Rust staining from this type of inclusion is clearly not the *result* of reinforcement corrosion, but such inclusions can be eroded quite quickly and can become a *cause* of corrosion (see Figure 3.7).

3.2.4 Pitting corrosion

Carbonation affects all concrete, although fortunately in most cases too slowly to reduce the **service life** of structures. **Chloride contamination** is less common, but usually causes severe damage quite quickly except in **high-quality concrete** or under dry conditions (see Section 3.2.1). Cracking, **delamination** and **spalling** can all be caused by chloride contamination and except in wet conditions the early signs are usually difficult to distinguish from those caused by **uniform rusting** due to carbonation. In wet concrete and when corrosion is more advanced, **pitting** may be visible enough to indicate that chloride contamination is the likely cause.

Making spot measurements of **surface electrode potential**, or better, mapping surface electrode potential over an area, can help to indicate whether chloride contamination is likely to be contributing to any corrosion that is found, but the only way to be sure is by the **chemical analysis** of samples. The investigation required is more appropriate to detailed investigations and Chapter 4 discusses sources of chloride contamination and ways of finding it in more detail.

Where enough **oxygen** is available, red rust (hydrated iron oxides and hydroxides) will grow with sufficient force to disrupt the concrete in much the same way as with general corrosion.

If chloride contamination is present and there is little or no oxygen at the corroding area (the **anode**), the corrosion will be being driven by a **cathode** with free oxygen access that is some distance away. This can happen in members that are only partly submerged in **sea-water**, the **corrosion products** at the anode then being **soluble iron compounds** which seep through the pores of the concrete or through fine cracks, with the result that the reinforcement may eventually disappear without leaving any trace of damage on the surface. When this happens, the corrosion is usually discovered only if the concrete is broken out or cored for some other reason; there are no tell-tale signs of it on the surface. If the corroding area is broken open and exposed to air, the corrosion products first turn black and then rusty in a matter of minutes. This type of corrosion is sometimes called '**black rust**' (see Chapter 4).

If chloride contamination is present in **saturated concrete** where the entire unit has little or no oxygen access, corrosion probably will not occur at all because there will be no active cathode. Corrosion is then said to be under **cathodic control** and the potentially cathodic area to be **polarized**.

3.3 Reporting the investigation

Investigating a structure for damage can make the investigator vulnerable to claims of **negligence**. The investigation must of course be made with appropriate care, knowledge and thoroughness even if the client has asked for no more than 'a quick cheap visual inspection', at least initially. To minimize the investigator's risk it is important that a brief for the work is agreed with the client, that the report sticks strictly to the **brief**, and that any content which is not fact is clearly shown to be opinion. This need for caution is true of most inspections, but it can be especially important in the case of reinforcement corrosion which can be very difficult to spot in its early stages.

When investigations are made for colleagues in the same organization or clients with a working knowledge of the relevant technology, for example other consultants, highway or airport authorities or in some cases owners of industrial complexes, it may be possible to agree the brief in enough detail to avoid most risks, but investigators must bear in mind that it will usually be the client's **insurers**, not the client with whom the investigator may have a good relationship, who will initiate any action against the investigator if there was a failure to report a problem which later causes the client loss.

> *Warning*. Beware of what you report, especially from a quick preliminary inspection, unless you are sure that you have your risks under control.

The **report** at any stage of assessment should record the findings strictly in relation to the brief: a well-drafted brief can already contain most of the structure for the report and save a lot of time at the report-writing stage (see 'Checklist of briefs for investigations').

The report must record what inspections were made, when they were made, and who was present (see the 'Checklist for reports'). It is highly desirable that the client should be present for at least part of any **initial investigation** and the report should record any important information exchanged during such visits.

Many organizations have a standard format for this type of report, but it is worth emphasizing the importance of recording relevant dates. They can be decisive in this work because years sometimes pass before action is taken following a report, and damage can increase sufficiently in this time to render the report invalid.

The report of an initial investigation must record what defects and indications of deterioration caused by corrosion were found in the (usually limited) areas that were examined. The implications of these observations should be explained and where appropriate recommendations should be made, but inferences and opinion should be clearly distinguished from facts. It is important to remember that it is much easier to be confident about a problem that has been discovered than to be sure that none exists, especially where chloride contamination is involved. If no damage was found in an initial investigation the report should therefore outline what further steps should be taken to determine whether deterioration caused by corrosion can be expected in a given period in the future, but a report which recommends *nothing* more than **monitoring and re-inspection** at some future date can miss an important opportunity to take **preventive action** before deterioration becomes significantly more difficult and expensive to repair. Reports of later stages of investigation will be detailed enough to specify what types and quantities of repair or protection are needed.

Tip. Try to ensure that the client is present for at least part of the time the structure is inspected.

Tip. Although the report of the initial investigation may not be intended for use as **evidence** in a **dispute**, if a dispute should develop it is important that the report contains nothing stated as fact that might be contradicted in a later more detailed report.

3.4 Initial and strategic decisions

The **initial investigation** should give enough information to allow the **client** to make sound **decisions** on what to do next. What clients choose to do will depend on **economic and strategic issues** that are important to them, at least as much as on the technical facts. For example, if the investigation was commissioned with a view to buying a structure, the client can look elsewhere or should at least be advised to commission a very thorough investigation which might well include looking for evidence of recent inspections of the same structure (why does the vendor want to sell?). On the other hand, if the client already owns the structure and hopes to continue to use it, there are **options** which the investigator can help the client to evaluate. These range from doing nothing, apart from carrying out regular **monitoring and safety** inspections, through protection and repair to prevent the condition of the structure from getting worse, to **improvement and upgrading**. An important additional consideration is often whether the client has grounds to take action against those responsible for causing the loss.

The initial investigation should have made it possible to estimate what further **detailed investigation** will be needed to establish the cost of competing options, but if for some reason it does not, a little further investigation and testing may be recommended as an intermediate step. Chapter 4 can be used for guidance on what might be done.

4. Detailed investigations

Chapter 3 gave general information about investigations, how they should be set up, carried out and reported. This information is also relevant when carrying out detailed investigations but is not repeated here. Chapter 2 should also be read as an introduction to this chapter.

Detailed investigations are normally undertaken by suitably qualified and experienced people. Unlike the preceding chapter on initial investigations, which is intended to guide engineers in general practice who may not have expertise in this area, this chapter accepts that many details of the operations will already be familiar to those who carry out investigations. This chapter is intended to give them additional information based on a wide consensus of opinion and experience, to provide basic guidance for those who wish to develop a special interest in this type of work, and especially to inform those who commission specialist investigations and rely on their results. The requirements for a detailed investigation should result from discussion between such specialists and the client, and this type of investigation is normally carried out to a detailed **specification** under a formal, although flexible, **contract**.

Chapter 5 which follows outlines repair options and methods that will affect the information needed from the detailed investigation. Chapter 5 should therefore be considered together with this chapter.

4.1 The purpose of detailed investigation and testing

When a detailed investigation has been completed and **reported**, the client should be in a position to compare the cost and value of the **options**. These can include options other than carrying out repair, for example selling or redeveloping, but the purpose of the detailed investigation is itself to provide enough information for the client to estimate the comparative cost and value of the repair options.

The investigation should establish how far **deterioration** has progressed and identify the causes of corrosion in enough detail to include an estimate of what additional areas would be affected in the future and which protection and repair methods could be used to restore or preserve the structure. The investigation should also provide enough detail about the individual types of repair that could be used to enable the investigator, in conjunction with the client, to define and specify repairs sufficiently for contractors to tender, though it has to be accepted that exact quantities may not be calculable until repairs are under way.

Where an investigation is carried out knowing that it may be used to provide **evidence** in a **dispute**, the nature and causes of deterioration will have to be investigated and recorded in enough detail to withstand questioning and denial by opposing parties.

4.2 Drawings, accommodation, equipment and access

Investigation is quicker and more reliable if the design drawings and the history of construction of the structure are available. If as-built drawings are available which also record alterations that occurred when the structure was built, the records will be invaluable (and most unusual). If there is any possibility of structural inadequacy, drawings will help greatly in allowing the present structural condition to be assessed and compared with the **design assumptions**. Drawings of structures even a few years old are usually hard to track down, but sometimes they can save months of investigation.

Even in cases where the original design drawings are not needed for **structural assessment**, the investigator will have to make drawings in at least enough detail to **record the investigation and test results**. If the structure is large it is a great help if drawings of details include key or location drawings, and the addition of **photographs** and isometric or perspective sketches can save much written explanation.

For use as **evidence** in **disputes**, additional pictorial information added to drawings can be very helpful to people who may be unfamiliar with the conventions of engineering drawing.

On large and complicated investigations, the time invested in preparing drawings on computer-based systems will almost certainly be worthwhile because it is then quick and easy to edit drawings and add the information which is collected as the investigation proceeds and when repair is specified. Computer-based drawings also have the advantage that they can easily be shared and adapted quickly in a variety of different ways to suit the particular points being illustrated.

Initial investigations are usually made without much more in the way of accommodation and facilities than the boot of a car, common tools, a **cover meter**, a **half-cell** and a **phenolphthalein** spray. More **detailed investigation** of even quite a modest structure needs access to a range of equipment, often including power supplies, and can generate a lot of paper work (see 'Checklist of useful equipment for investigations' at the end of this chapter).

The investigation will be quicker and cheaper if adequate secure accommodation is provided for personnel and equipment on the site itself. Accommodation is one of the few things that is normally easier to arrange in investigations of existing structures than for new construction: there are few existing sites where a co-operative **client** cannot make reasonable accommodation available for the required period and it is worth making an effort to arrange it.

Investigations nearly always have to be carried out on structures that are in use. In the case of **occupied buildings**, the people who occupy them may be trying to work, learn or sleep while the work is going on.

Investigators should carry and offer identification and try to introduce themselves to any representatives of the **occupiers**. A little trouble taken to consult and inform the occupiers about what is going to happen to their environment, when and for how long, can work wonders and prevent the endless aggravation and interference which can make working on occupied sites an expensive nightmare.

The cost of **access equipment** for survey and testing operations can be high. In cases where it is already known that repair is inevitable, there are many arguments for postponing as much of the detailed investigation as possible until access and equipment for repair operations are in place, but at least in the early stages of the work, some dedicated access equipment will usually be needed. This can range from ladders, through **abseiling equipment**, to cradles, hydraulic platforms and specialized access platforms that can reach under high bridges or into other awkward places. Sometimes it may be economic at the investigation stage to provide temporary works which can remain in place for the repair phase, for example scaffolding or fans and netting to catch falling pieces of concrete. If the repair phase is to be done when ambient weather conditions will be unsuitable, protection can be put in place before the investigation starts.

The testing to be done will influence what access equipment is needed and this must be thought through before the access equipment is chosen. It is quite difficult to use even a simple cover meter or SLR camera adequately from a ladder (or when hanging from a rope), and impossible to break out concrete except in very small areas, or to cut cores of any kind, without a reasonably stable platform.

Where **evidence** is being collected for a **dispute**, it can be helpful to allow the opposing parties the use of any fixed access equipment so that they can collect and agree their own evidence.

4.3 Recording defects and test results

It takes a surprising amount of time to keep good **records** on site, but properly made records can be used with little further effort to relate **deterioration** to the design of members, the position of underlying reinforcement and the local environment. Later they can be used as the basis of the report and later still to define the work to be done by the repair contractor, and finally, to record what he has actually done. In the eagerness to get started on an investigation spending time on thinking through how the results are best recorded for easy use later seems an unnecessary hold up. To coin an old proverb about marriage, 'report in haste, repent at leisure'. Records of repairs to structures should form part of a **'Log Book'** to be kept by the **client** for reference in later years when work might be needed to deal with further deterioration or to prepare for a change of use.

Sketches and photographs should be made of all areas with significant visible **defects** such as **cracks**, **spalls**, **rust stains**, **honeycombing**, etc. unless the structure is very large when examples of typical defects usually have to suffice.

The ends of significant cracks should be permanently marked on the concrete and the lengths between end marks, and the approximate **widths of cracks** recorded. The depth of reinforcement found at spalls, and the condition of any reinforcement which is visible, can be significant for assessment and should also be recorded (see Figures 4.1–4.3).

Film is cheap compared to an investigator's time (magnetic storage even cheaper) and **photographs** can be a great help in recording inspection details provided that the location of each photograph is properly identified. Probably the best way of recording

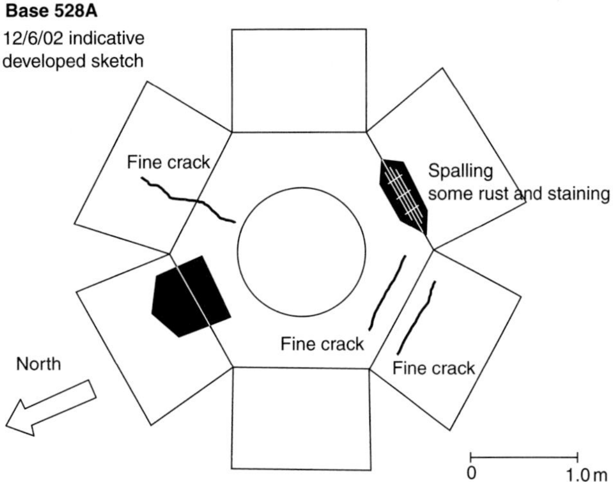

Figure 4.1 A developed view recording damage at a hexagonal plant base at a refinery.

the location of a photograph is to attach a note of location details (and date) to the structure so that this information becomes a permanent part of each photograph. When dozens of very similar portraits of spalling concrete have to be sifted weeks after the inspection, this type of record removes all doubts about what it is one is looking at.

Resist the temptation always to take photographs very close to the defect in the hope of recording more detail: photographs with a wide enough view to show the defect in relation to its surroundings are not only more likely to be in focus but also often more useful than close-ups. If close-ups are needed, it is a good idea to take a second shot of the same area.

An SLR camera with a wide-range zoom lens, or interchangeable lenses is very versatile, but when access is difficult or there is a lot to carry, a compact camera which can be operated with one hand while the other clings onto a ladder or a swaying cradle tends to get used a lot more often. If it is worth taking an SLR, it is probably also worth taking a tripod so that photographs can be taken without flash even in a poor light: this is especially useful for photographing **core** holes.

4.3.1 Sampling

Choosing where to make measurements or take **samples** is one of the most important decisions that the investigator has to make. For **initial investigations**, measurements

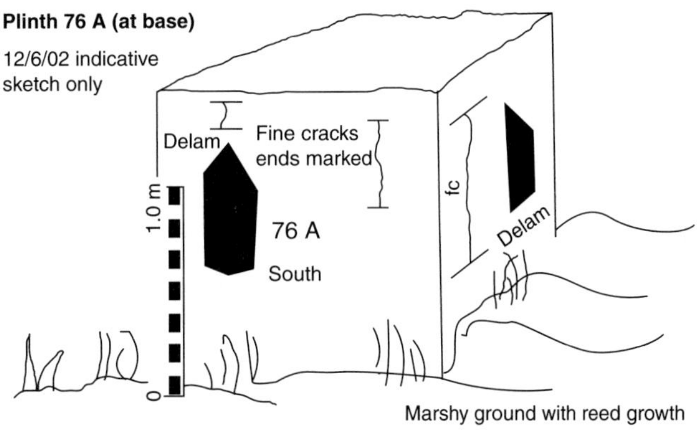

Figure 4.2 A perspective sketch recording damage at a square plinth.

Initial survey

Typical sketch from an initial inspection of a 13-year-old 1960s RC framed office building

The survey was made by a non-specialist engineer working from a cradle

P17 etc identify photographs taken with compact zoom databack camera

Spot covermeter readings are marked with values referring to the front face unless indicated otherwise by "soffit" or "s" for side

Covermeter was set assuming 12.5 mm dia for links

Wisely, no attempt was made to remove concrete at this stage

After a fully detailed inspection, the client decided that repair and protection was the best option

Areas where reinforcement was in or very close to carbonated concrete were repaired with replacement concrete. Other areas were coated with anti-carbonation coatings to reduce further carbon dioxide penetration

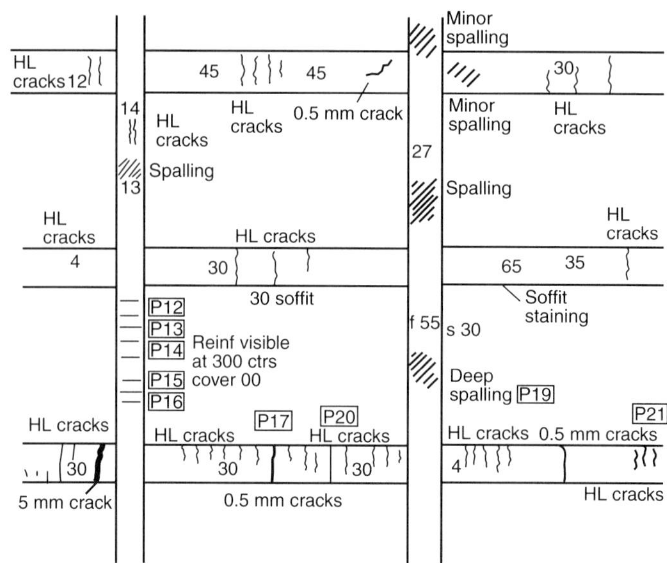

Figure 4.3

are usually made in typical or worst places that already show signs of damage or places where **deterioration** is most likely to result in damage in the future. **Detailed investigations** intended to quantify existing damage and predict future damage in the whole of the structure should be organized on a **statistically valid** basis that allows general inferences to be made from limited data.

Factors to bear in mind when planning **sampling schemes** relate to the investigator's knowledge of how the structure was built. For example, the **concrete quality** will be much the same throughout one **pour** apart from the possible effects of settlement near the top of the pour and **honeycombing** from leaking **formwork** joints near the bottom, but can vary significantly from one pour to another. Poor quality aggregates or damaging **admixtures** like **calcium chloride** are also unlikely to vary except between pours.

Where the **depth of cover** is sampled, workmanship is likely to have been critical. Fixing of reinforcement and the use (or absence) of **spacers** is likely to be similar in sections of similar shape and size, but can vary in different conditions.

If the structure was built under more than one **contract**, the sampling scheme should look for any systematic differences in workmanship and materials attributable to the different contracts.

Helpful general comment on sampling is given most textbooks on statistics and specific advice is given in the BS 2846 Series [5] and BS 6000 [6].

4.4 Cracking, delamination and spalling

These defects are discussed in Sections 3.2.1 and 3.2.2. In a **detailed investigation** there is little to do that is different from an initial investigation apart from making use of the opportunities to break out cover at representative locations to allow the condition of the reinforcement to be inspected and the **depth of carbonation** to be measured with the **phenolphthalein** test (see Section 4.6.1). Carbon dioxide penetrates relatively quickly down cracks and carbonation depths are likely to be greater in the neighbourhood of cracks. Examination here may give useful information on the danger that lower layers of reinforcement are in carbonated concrete, especially significant if the lower reinforcement is parallel with what appeared to be a harmless transverse crack at the surface. If **electrochemical protection and repair methods** are among the **options** that will be considered, physical damage (other than the removal of delaminated concrete) should be kept to a minimum during investigation to avoid unnecessary repairs.

Active cracks are wide transverse cracks that continue to move with changes in load or temperature. They are not controlled by the reinforcement. Active cracks behave as informal **joints** and cannot be repaired by injecting or sealing, but must be repaired by treating them as joints except in the rare cases where the cause of movement can be corrected (see Figure 4.4). It is therefore necessary to identify them in the investigation. Where it is suspected that a crack may be acting as a joint, the crack width should be measured at intervals of 2–4 hours, preferably with a demountable **strain gauge** to discover whether or not the crack is active. If a strain gauge is not available, a vernier calliper measuring between two small cross-head screws glued into holes drilled in the concrete is repeatable enough to detect the most significant movements if the same calliper is used. Stainless steel screws can be left in place for checking at a later date.

Delamination is not usually visible on the surface but can be found by hammer-tapping. The approximate boundaries of hollow-sounding areas should be marked on the structure and then **photographed**, or sketched on the drawings. It takes a little practice to be sure that a hollow sound means delamination and if the delaminated part is more than about 50 mm thick and the void small, delamination can be difficult or impossible to identify.

In enclosed places that echo, and on noisy sites, it can be very difficult to detect the sound of delamination. Working at night, when plant is **shut down** or road traffic is light, can help, but on some sites soundings are impossible. A thermal-imaging camera is an expensive but useful, instrument which will identify delamination if it is used when the ambient temperature has changed quickly, for example early on a sunny morning. It has also been suggested as a means of locating changes in moisture content, for example at leaks in membranes or embedded services [7].

Areas of delamination on horizontal slabs can sometimes be seen if dry dust or fine sand is scattered on the surface. When the surface is tapped with a (heavy) hammer, the dust over the delamination bounces, whereas the dust over solid concrete is not disturbed.

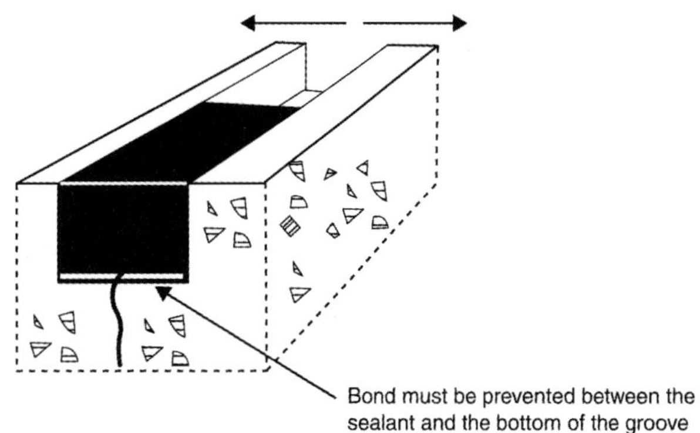

Figure 4.4 Active cracks behave like joints: they must be made into sealed joints.

Suspected delamination can be confirmed by hammering until pieces come away. Although this creates the additional problems of removing the debris and making good the damage, at the stage of detailed investigation it can be helpful and it avoids these areas being included in **electrode potential mapping** surveys or areas treated by electrochemical methods of protection and repair.

> *Warning*. Spalling concrete can kill – delamination should be treated with great respect: hard hammering can bring the cover down in pieces big enough to cause serious injury.

4.5 Surface electrode potential measurements

4.5.1 General principles

The electrochemical nature of reinforcement corrosion (see Section 2.1) makes it possible to use electrical measurements to study the corrosion process. Some measurements can only be made in a laboratory under artificial conditions, but **electrode potentials** can easily be measured and mapped on the surface of real concrete under site conditions. These potentials are related to the condition of the steel in the concrete and can be used to distinguish between steel that is **passive** and steel which has lost its protection and is in a condition to corrode, i.e. **active**. The measurements are extremely useful, but they do not show how quickly the steel is corroding, and in some cases 'active' steel may be corroding at a rate that is too slow to matter.

One of the most useful applications of surface electrode potential measurement is to pinpoint where **chlorides may have penetrated** to the reinforcement and it can be used as a quick way of locating the most important points to take samples for **chemical analysis**. On very wet concrete surface electrode potential measurement can sometimes also be used to pinpoint **anodic areas** where reinforcement may be corroding unseen. There will usually be no cracking even if corrosion is very serious.

Experts disagree about the reliability and interpretation of *individual* surface potential values as a diagnostic test of corrosion, but they do mostly agree that comparing results, for example by plotting a contour map of equal potentials, can identify the places where corrosion is possible because of a loss of passivity.

4.5.2 Half-cell measurements

The '**half-cells**' which are the basis of surface **electrode potential measurement** get their name from the fact that one half of an electric cell is formed by the reinforcement in the concrete and the half-cell probe used on the concrete surface is the other half-cell.

Many materials can be used to make a half-cell, but the principle is always that the half-cell consists of an **electrode** in a solution of known concentration of a salt of its own **ions**. The practical combinations that are most often used are copper in copper sulphate solution and silver in fused or electrodeposited solid **silver chloride**. Mercury with calomel (mercury chloride) is used as a **reference electrode** for checking practical half-cells, and the standard reference potential from which potentials are measured is the standard hydrogen electrode (see Section 4.5.4).

Copper/copper sulphate cells are widely used for site measurements, but where it is necessary to embed a half-cell permanently in concrete, for example for long-term **monitoring** or to control cathodic protection, **silver/silver chloride cells** are normally used because they have quite good long-term stability. Silver/silver chloride cells based on saturated potassium chloride give potentials 98 mV less negative than copper/copper sulphate cells.

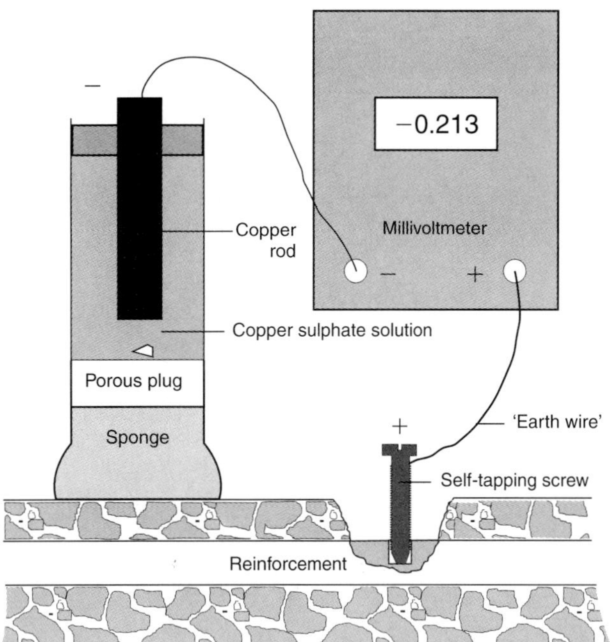

Figure 4.5 Half-cell probe measuring surface electrode potential.

Contact between the concrete and the **electrolyte** in the half-cell is made through a sponge wetted with conductive water. A little detergent makes tap water sufficiently conductive.

To make measurements the half-cell is connected to one terminal of a high impedance voltmeter and the reinforcement is connected to the other. Because the **corrosion currents** flowing in the concrete are very small, the voltmeter should have an input impedance of at least 10 megohms to avoid it short-circuiting the corrosion cell. Most general-purpose digital multimeters have a sufficiently high impedance. In very dry concrete, the impedance may need to be 1000 megohms or more, though surface electrode potential measurement is anyway difficult to carry out in such conditions (see Figure 4.5).

4.5.3 Connection with the reinforcement

As so often happens, it is the apparently simple practical details that need most attention; in this case it is the electrical connection with the reinforcement. A hole has to be cut to reach the reinforcement and a smaller hole must then be **drilled into the reinforcement** to make a good electrical connection. A self-tapping **connecting screw** is the usual way of making a connection that is robust enough for most site conditions. A suitable reinforcing bar has to be found by searching carefully with a **cover meter**. The bars nearest the surface are usually 'links' but it is often easier to make the connection to a deeper bar of substantial size. If only a limited area is being investigated, it is best to choose a bar in or reasonably near it.

The **depth of cover** determines the area of concrete that needs to be cut away to expose the bar because there must be enough room to get the chuck of an electric drill very close to the reinforcement to drill the connection hole.

Normally, a diameter of at least 50 mm is cut away, either with a hand-held coring bit or by cutting between a series of drilled holes with a cold chisel. Holes joined by cutting with a cold chisel can also be used to measure the **depth of carbonation**

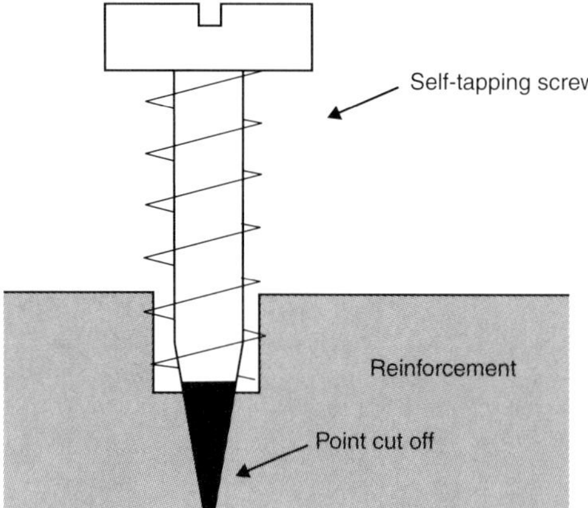

Figure 4.6 Cutting off the point of the connecting screw allows the use of a shallower hole.

(see Section 4.6.1). Drilling reinforcing steel is difficult and time can be saved by using a special steel bit (for example a **Cintride-tipped bit**). A short length bit (about 50 mm) makes it possible to use considerable force without breaking it. On large sites where many connections have to be made, it can be helpful to tap the drilled holes with a thread (see Figure 4.6). If connections are to be left in place for later measurements, stainless steel screws or wire tails can be left protruding from the back-filled connection hole. On accessible structures such connections are inconspicuous and less likely to interest vandals than expensive junction boxes.

A robust insulated flexible copper wire is connected to the earthing point and tied to secure points at intervals along its route to prevent the accidental loss of the connection. How far it is possible to work from a single 'earth' connection depends mostly on how well the bars are inter-connected. The total circuit resistance should not be much more than 1 ohm (including the connecting cable), and this has to be tested by exposing some reinforcement near the measuring point and measuring the resistance between it and the connecting screw. These temporary connections do not have to be robust, and provided the steel can be cleaned (for example by the bit drilling down to it), there is no need to drill a separate hole or use a self-tapping screw.

4.5.4 Checking the half-cell

Before taking any readings, it is important to check that the **half-cell** is giving results that have meaning. Two half-cells can be checked against each other by connecting one cell to each terminal of the millivoltmeter and touching the sponges together. If the readings are exactly the same they cancel out, so the reading on the meter is zero. In practice, a difference of up to 20 mV is reasonable. Some users hold the bare copper wire of the 'earth' connection between the fingers or in the mouth and touch the sponge just to check that there is a reading of some sort (−50 to −200 mV is typical) and everything is working.

There is very little to go wrong with the **copper/copper sulphate half-cell** other than the sponge being too dry or badly contaminated with copper sulphate; the copper sulphate solution being too weak or too little to reach the electrode; or the copper electrode being contaminated. The remedies are obvious.

In the UK, half-cell readings out of doors are normally between −50 and −300 mV where the steel is passive and can be as low as −600 mV where it is **active**, especially under damp conditions. If readings are not as expected, or if they are erratic, there may be a source of error.

Half-cell readings become more negative with increasing moisture content. If comparisons are being made between one point and another over a short period of time (say in mapping) this does not matter because the pattern of readings will not be affected, but readings taken where there are visible differences (for example wet, repaired or delaminated areas) can give misleading results and should be avoided or at least the differences recorded with the results. The numerical values for different areas surveyed at different times cannot be compared, nor should it be expected that the same values will be found if a previously surveyed area is surveyed again when the conditions are different.

Rendering, paint and surface contamination can make readings less negative, especially if the reinforcement is **passive**. Visible coatings can be removed with a dry sanding disk if they cannot be avoided. **Silane** treatment on bridges can make surface electrode potential measurement very difficult, and sheltered areas under bridges sometimes develop a thin skin of **carbonation** which can make the readings positive if the steel is passive. Abrading up to 2 mm of the surface usually cures the problem.

If the concrete is interrupted by significant **delamination**, or even **cracking**, readings can be more positive because of the long path between the half-cell and the reinforcement. It is best to avoid such areas.

4.5.5 Taking and interpreting half-cell readings

Readings are taken by touching the concrete surface with the sponge and the values obtained should be reasonably steady for several seconds. A slow count from 0 to 10 should show the values drifting by less than 10 mV. Faster drift is usually the result of dry concrete taking up water from the sponge and this can be cured by spraying the whole of the surface lightly with water and allowing the moisture to soak in for an hour or two before readings are taken. A surface that is running with water also gives unreliable results.

By an inconvenient convention, surface **electrode potential** measurements are expressed as the potential of the reinforcement relative to the half-cell, so the reinforcement is connected to the positive terminal of the meter and values will normally be negative. The terms 'more than' or 'less than' are therefore ambiguous and 'more positive than' or 'more negative than' have to be used instead.

The **'Van Daveer' criteria**, which are accepted by ASTM, suggest that the values give a direct indication of the likelihood that **anodic** areas are corroding. Table 4.1 shows the values for a copper/copper sulphate electrode.

As readings are affected by the dryness of the concrete and the temperature, many experts reject this simple interpretation though experience shows that it usually about right, except sometimes in **saturated concrete** where values more negative than −600 mV can be found in concrete where there is insufficient **oxygen** to sustain corrosion (see Section 1.2.1).

Table 4.1 Probability of corrosion for a ***copper/copper sulphate half-cell***

Potential	Probability of corrosion
More positive than −200 mV	<10%
More negative than −350 mV	>90%
Intermediate	Uncertain

4.5.6 Electrode potential mapping

Using **electrode potential** measurements to give relative rather than absolute data is a way around most of the problems of interpretation and lack of repeatability. By drawing maps of points of equal potential on the surface of the concrete the investigator can consistently locate areas where **passivity** has been lost, in spite of the fact that from day-to-day individual values may vary by more than 200 mV, depending on the weather (see Figure 4.7).

For surface electrode potential mapping, grids are drawn on the concrete and readings are taken at the intersections of grid lines. Sampling for the possible presence of **chloride contamination** can be made in limited areas where the contours show activity to peak.

Establishing a tidy and reasonably accurate grid takes time and trouble. If the surface of the concrete is very dry and likely to need to be wetted to produce reliable readings, it helps to draw the grid and add any location data while it is dry. Wax crayons are often used, but ordinary blackboard chalk, perhaps coloured, can be excellent for markings if they do not need to last for very long.

If the grid is likely to be reused at a later date it must be accurately drawn in a position where it can be reproduced, permanently marking perhaps the top left-hand corner (as in a spreadsheet). The position of the grid should also be marked on drawings to save a

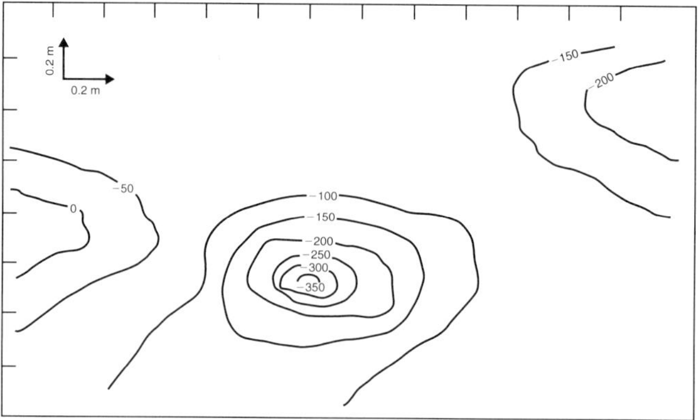

Isopotential map under dry conditions

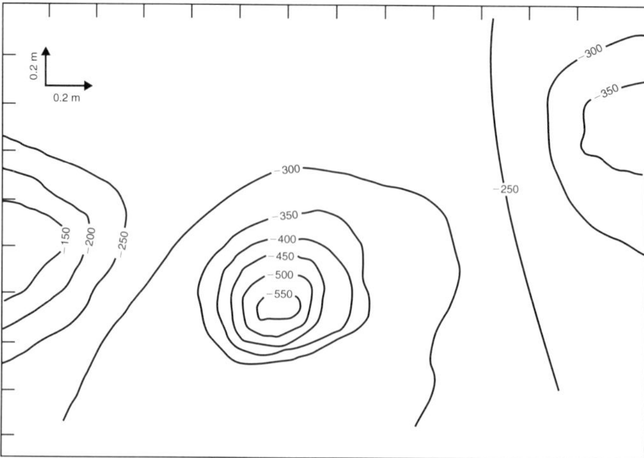

Isopotential map under wet conditions

Figure 4.7 Surface electrode potential mapping pinpoints the danger areas, whatever the conditions.

lot of detailed work in preparing records, and photographs of the grid position and markings can be a great help in refreshing the memory about surveys which were carried out long ago or a long way from home. To have the values of the readings *legibly* marked on the concrete makes **photographs** valuable in combating lapses of memory.

Electrode potential maps can locate areas of lost passivity, but not all of these necessarily contain steel that is corroding. If corrosion is taking place in an area of relatively low potential, is quite possible that it is providing **sacrificial protection** for neighbouring areas where the potential is only a little higher. In such cases, repair of the worst-affected area would end this protection and expose the neighbouring areas to the risk of corrosion. Electrode potential mapping is especially useful for identifying this risk and discouraging ill-considered local repair that could soon lead to problems in adjacent areas (Section 2.2.5 briefly explains the mechanism and reference [8] records what is probably the earliest real understanding of the problem).

Logging half-cells are available with microprocessors to plot electrode potential maps automatically or to download them to a spreadsheet. They greatly speed-up the process of collecting the data, but as with all automatic data logging, have the danger that unusual values will not be noticed while the operator is on site and able to investigate them.

Helpful information on mapping is given in reference [9], and reference [10] explains the theory and practice.

4.6 Measuring carbonation and cover depth

4.6.1 Carbonation

All concrete is affected by **carbonation**, which results from carbon dioxide slowly penetrating the concrete cover, reducing the protective **alkalinity** as it goes. Sooner or later the **carbonation front** will reach the reinforcement in any structure and leave it unprotected from **rusting** (see Section 2.2.1). Designers cope with this natural ageing process by trying to ensure that the penetration rate will be slow enough and the cover thick enough to protect the reinforcement for the planned **service life** of the structure (see Section 4.6.3).

Lack of cover quality or quantity (often both) allows the carbon dioxide penetration to reach the reinforcement sooner than intended. This usually results from bad practice during construction and many experienced investigators agree that they seldom seen defects caused by corrosion from carbonation where the cover **depth of cover** and **concrete quality** comply with any reasonable specification.

The time when the reinforcement will be at risk of rusting can be calculated from the rate of penetration of the carbonation front and the thickness of cover. In the existing concrete the rate penetration of the carbonation front is calculated from the formula $d = K\sqrt{t}$, where d is the depth of the carbonated concrete in millimetres, t the time since construction in years, and K is the permeability constant of the concrete, usually called the '**carbonation coefficient**' in this context.

If t is known and d is measured, K can be calculated and then used to estimate what d will be at any time in the future. The carbonation coefficient K is an inbuilt 'hallmark' of the quality of the concrete. **K-values** range from about 0.5 to 3 for precast concrete and from about 1 to 9 for **in-situ concrete**, with an average of about 5 [11]. (Some experts argue that the depth of carbonation increases only approximately in proportion to the square root of the age of the concrete and that K is an unreliable measure, but for the practical purpose of assessment the relationship is generally accepted as being at least as accurate as any other part of the process.)

The **depth of carbonation** *d* is quite easy to measure by breaking away pieces of concrete at representative places and spraying the freshly exposed surface with **phenolphthalein indicator solution**. Concrete which is uncarbonated will show a bright pink stain which persists for several hours or even days after exposure.

Carbonated concrete will not produce a bright pink stain, but in some types of concrete, a paler pink coloration may initially cover the surface even if the concrete is carbonated. Concrete containing **limestone** fines, and mortars containing **lime** are examples. This pale coloration is quite distinct from the coloration given by uncarbonated concrete and usually fades within an hour or so of exposure.

Because it is so easy to do, the phenolphthalein test is often applied rather carelessly and then gives disappointing results. The most common problem is the tendency to spray on so much indicator solution that it runs over the surface and spreads the pink stain over areas that should be colourless (see Figure 4.8). A good-quality adjustable sprayer which allows the solution to land on the surface as a fine mist helps to avoid this, but if excess solution has run, waiting for a few hours before trying to measure the depth often makes the separation distinct. Except in concrete with a very high cement content (say more than about 350 kg/m^3), the colour fades in a few days as the exposed surface becomes carbonated. It is important to make the carbonation test on a broken, not a cut or drilled surface, because cutting usually exposes some **unhydrated cement particles** and gives a smudged or optimistic picture (see Figure 4.9).

Phenolphthalein indicator solution is a commonly used laboratory reagent and is normally supplied as a 1% solution of (solid) phenolphthalein in ethanol. For use on dry concrete water must be added to ensure that the alkaline constituents of the cement are in solution. Between 10% and 50% water gives satisfactory results, but under very dry

Figure 4.8 Marking location details on the concrete makes this photograph a permanent record of carbonation and cover depths.

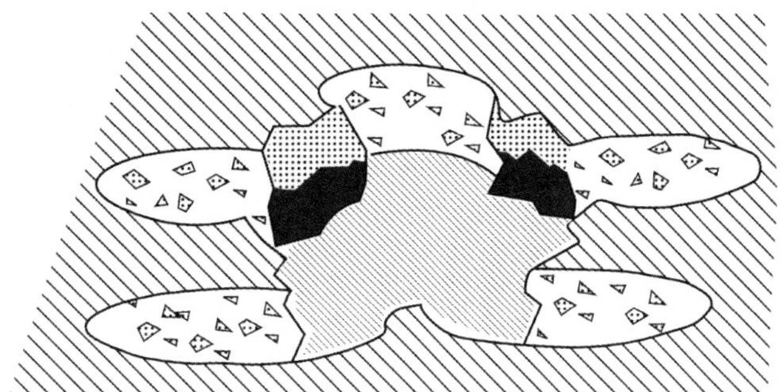

Figure 4.9 Drilling a series of large holes helps to break out pieces for the phenolphthalein test.

conditions it can be helpful to spray the surface with water before testing with phenolphthalein solution. Too much dilution will precipitate phenolphthalein which is not soluble in water alone. Although ethanol or industrial alcohol are usually prescribed as the solvent for phenolphthalein and deionized water as the 'diluent' for the solution, for use on site methylated spirit and tap-water are equally satisfactory. In the unlikely event that an investigator in a remote location has phenolphthalein powder, but no industrial alcohol, gin or any sugar-free beverage containing close to 50% alcohol will dissolve a sufficient concentration of phenolphthalein for emergency use.

Experts point out that the level of **alkalinity (pH)** at which phenolphthalein changes colour (about pH 9) is slightly lower than the alkalinity of uncarbonated concrete (more than pH 11) and that the test therefore gives an optimistic result. In practice the small difference is insignificant in a measurement that cannot be made with great precision when making assessments.

> *Tip.* Textbooks say the solution should be freshly made. This is not strictly necessary, but if the indicator solution has been stored for many months (in a stoppered container), check that it turns pink on an alkaline surface (for example cement, soap or household scouring agent) before using it.

> *Warning.* Phenolphthalein should be treated with respect: it is a powerful laxative and may also have longer-term detrimental health effects.

Reference [12] gives helpful information.

4.6.2 Measuring depth of cover

Unless cracking over the reinforcement or **spalling** are found, there may be no visible evidence to indicate an insufficient depth of cover. Estimating the depth of cover where there are no signs of damage is important in the investigation because it allows the investigator to predict when and where the reinforcement will be at risk from **general corrosion** at a later date.

Cover depth can be estimated non-destructively by using an electromagnetic **cover meter**. Cover meters are affected by the size as well as the distance of the reinforcement and for accurate measurement simple cover meters have to be set to the appropriate bar size: some advanced cover meters may calculate this automatically. A useful, if only approximate, initial estimate of cover depth is in any case simple to make by setting the meter for **reinforcement of an assumed bar diameter**. Potential trouble spots can be located quickly by sweeping the surface with the meter preset to sound a warning when the cover is less than say 40 mm assuming a 12 mm diameter bar.

More precise estimations of cover depth with a cover meter take *much* more care and even with the greatest care and advanced cover meters, on site it is not possible to estimate more closely than about ±5 mm. This is precise enough for most practical purposes considering the variable nature of the data, but if confirmation of the depth is needed (for example, when evidence is being collected for a dispute) the depth of representative bars must be found by drilling.

Bunched bars and bars that cross one another can confuse cover meters as can magnetic **aggregates**, but the cover meter remains an excellent, and indeed essential, survey tool. Even simple 'rebar locators' which are not designed as precision cover meters can give a useful indication of cover depth over the outermost layer of bars as well as precisely locating their positions. From time to time throughout any large inspection the cover meter should be calibrated against the measured distance to the reinforcement by drilling to confirm its position or by measuring the bar size at a convenient spall. A vernier calliper is a convenient tool for measuring depth and can also be used to estimate crack movement and as a comparator when estimating crack width.

Tip. Drilling good-quality concrete (for any purpose) is easiest with large bit sizes in heavy-duty drills. Bits of at least 20 mm diameter are best and holes of less than this diameter are difficult to repair with cement mortar when the investigation is finished. The more convenient cordless tools with smaller diameter bits have only limited uses on concrete.

Cover meter readings are easy to challenge in **disputes** because several variables can affect the readings. Holes that allow the position of the cover to be measured are essential where **evidence** is being collected and should be left for inspection by opposing parties so that depths can be agreed.

4.6.3 Interpretation of carbonation and cover depth results

As well as being used to make **depth of cover measurements** at suspected trouble spots, a **cover meter** can give a useful indication of the general adequacy of cover over a large area if enough values are measured at predetermined locations which avoid operator bias. Cover depths plotted as **cumulative frequency diagrams** allow an estimate to be made of the percentage of cover that has less than a certain value. This can then be compared with **depths of carbonation** and **K-values** for corresponding areas to allow the time to corrosion to be estimated. Figures 4.10–4.12 show how the data are handled and how the 'hallmarks' are revealed. Initially, different parts of structures should be treated as separate populations for statistical purposes, but results from **samples** can be combined if the results show that this is justified. A histogram or **bar chart** of the results can give a useful check of whether the results are part of more than one **statistical** population. Reference [13] explains the prediction method in more detail.

For this kind of work the use of an advanced logging cover meter can save a great deal of tedium and expensive time by down-loading data and using a spreadsheet or special program to calculate and plot histograms, cumulative frequency diagrams and in some cases even to offer diagnoses automatically. However, all automatic data logging has the disadvantage that the significance of unusual values is usually not considered while the operator is on site and able to mark the spot on the concrete and investigate the

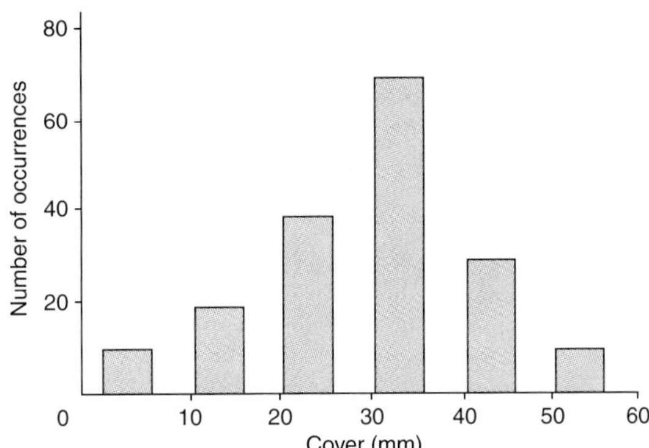

Figure 4.10 A histogram confirms that these results are part of a single population.

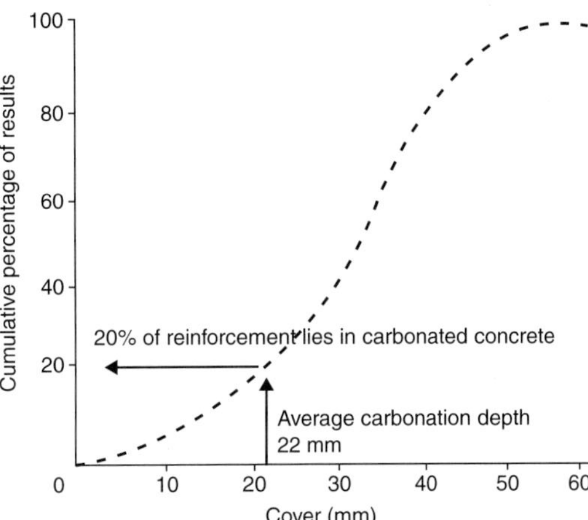

Figure 4.11 The cumulative frequency diagram indicates how much of the reinforcement has less than a given depth of cover. Adding the average depth of carbonation shows how much of the reinforcement is at risk of corrosion.

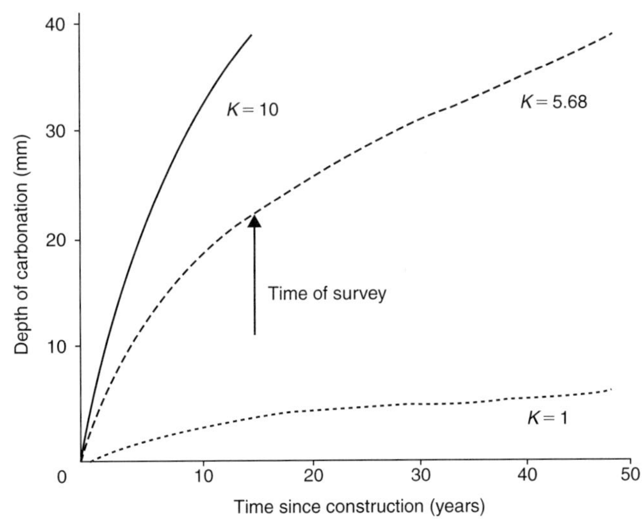

Figure 4.12 Plotting the K-value curve allows the amount of reinforcement that will be at risk of corrosion at any time to be estimated from Figure 4.14. For example, our sample building has an average carbonation depth of 22 mm 15 years after construction and therefore a carbonation coefficient (K value) of 5.68. From this we calculate that 30 years after construction the average carbonation depth will be about 32 mm and Figure 4.14 shows us that about 40% of the reinforcement will then be at risk from corrosion.

cause. Sometimes advanced 'intelligent' cover meters are advertised as 'thinking for themselves': perhaps they can, but thinking is the one part of an intelligent investigator's task never left to anything or anyone else.

Investigators inexperienced in using a number of different cover meters should arrange to try several before settling on one for any large investigation. The best suppliers normally welcome this interest. Useful information on cover meters is given in reference [7].

4.7 Chloride contamination

The presence of any **chloride contamination** at concentrations that can cause corrosion has a profound effect on the prospect of successful repair and can rule out most of the simpler repair **options**.

The presence, and more critically the absence, of a significant level of chloride contamination is a key factor in the assessment and can be confirmed only by **chemical analysis** of a sufficient number of **samples**. This is an expensive and time-consuming operation and is best preceded by identifying the most likely sources of chloride and where they might be found. Chlorides can enter concrete in two ways: they can be present in the materials when the concrete is mixed or they can penetrate the hardened concrete from the outside (see Figure 4.13).

As far as the possibility of **chlorides in the mix** is concerned, the location and date of construction of the structure can give clues as to whether or not unacceptably contaminated sea-dredged **aggregates** might have been used. As late as 1972 the permitted chloride ion concentration from marine aggregates was about 0.6% by mass of cement and this did not drop to (a presently acceptable) 0.35% until 1977 (both figures calculated from CP 110). In temperate climates where potable water is plentiful it was unusual for **sea-water** to be used as mixing water for reinforced concrete, but there is at least one recorded case on the south coast of England where the contractor found it simpler to let down a bucket on a rope than to walk to the nearest tap when building a small reinforced concrete marine structure. In temperate regions sea-water contains about 1.6% chloride ion and would therefore result in about 0.8% chloride in concrete with a water : cement ratio of 0.5. In tropical regions if tidal flow is restricted (for example, the Arabian Gulf region) the value can be nearly twice as high. In arid marine locations, sea-water has been used for major construction projects and produced recorded chloride concentrations of more than 1.3% by mass of cement, usually with unhappy results.

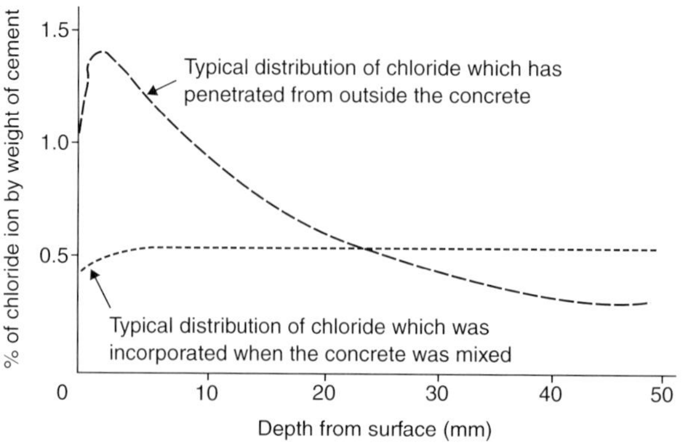

Chloride ion profiles in concrete from penetrated and mixed-in chlorides

Figure 4.13 Chemical analysis shows whether chlorides were in the mix or have penetrated later.

Another possible source of chlorides in the mix is **calcium chloride** accelerator. In the 1960s this was widely recommended for use in reinforced concrete at a concentration of up to 2% of anhydrous calcium chloride by mass of cement (1.3% as chloride ion): 'extra rapid hardening Portland cement' was sold with 2% of this admixture already in it and calcium chloride solution was delivered by tanker in a widely advertised product known as 'Hydrol'. In 1972 CP 110 restricted the calcium chloride content of reinforced concrete to a maximum of 1.5% by weight of cement (0.98% as chloride ion) which is well above the threshold for a high risk of corrosion (see Table 4.2 below) but the admixture was not specifically banned from use in reinforced concrete until 1985 (in BS 8110). The history of the construction of the structure and examination of weather records can indicate whether any parts might have been built during periods of cold weather when calcium chloride might have been used in **in-situ concrete**. Manufacturers of **precast concrete** elements used the admixture widely until the early 1970s, whatever the weather.

The possibility that **chlorides may have penetrated** the concrete from outside is easier to identify. Apart from wind-blown spray from nearby sea or brackish water, deliberately applied de-icing **salt** and accidental spillage of salty waste are the most obvious sources. Less obvious but not uncommon sources include wood-wool permanent formers and formwork (calcium chloride is used as a pre-treatment to prevent substances in the shredded wood from inhibiting the hydration of cement), spillage from lavatories in places where sea-water is (or was) used for flushing mains, and fire-fighting and even fire practice in coastal areas. Water is necessary for chloride contamination to enter concrete and regularly wetted areas are especially vulnerable, but wind-blown salt from the sea can accumulate in dry weather and be washed into the concrete when it rains.

A 'Checklist of vulnerable locations' is given at the end of this chapter.

Whatever the source, the high cost of testing makes it necessary to start with the areas where chloride contamination is most likely and the search can be narrowed down by the relatively quick method of **electrode potential mapping** (see Section 4.5.6).

Testing enough representative samples is the key to the successful estimation of the danger (see Section 4.3.1) but not all of the samples taken may need to be analysed. A large number of **samples** can be taken on site when it is convenient, concentrating on the areas pinpointed by observation and electrode potential mapping but including some areas where chloride contamination is less likely. Testing can be concentrated initially on a sampling plan where a selection of the these samples, say 1 in 10 or 1 in

Table 4.2 Likelihood of corrosion in uncarbonated[a] concrete (see Section 1.2.1)

Total chloride ion[b] by mass of cement (OPC[c])	Chloride in mix	Penetrated chloride
<0.2%	Very low	Very low
0.2–0.4%	Very low	Low
0.4–0.6%	Moderate	Moderate
>0.6%	High	High

[a] In carbonated concrete, much smaller amounts of chloride cause corrosion.
[b] Found by analysis after acid extraction.
[c] Cement containing less than 8% C_3A (e.g. BS sulphate-resisting or ASTM Type V) increases the likelihood of corrosion.

100, is chosen depending on the size of the job. Testing is then continued only until the consistency of the results shows that a representative picture of the chloride contamination has been produced. Because analysing **samples of drill dust** costs at least five times as much as the marginal cost of collecting additional samples when operatives are already on site, this plan can be quite economical, but it is especially useful if the investigator is working far from home and cannot conveniently revisit the site, or in **occupied** structures where there may be limited opportunities for the intrusive process of hammer-drilling concrete.

Samples can be taken by breaking off suitable pieces of concrete that have been carefully marked to record their location and orientation, by collecting the dust from drilled holes or from **cores**. Drilling for dust samples makes it relatively easy to determine how the chloride content varies with the depth from the surface and drill-dust samples need no further grinding before division and analysis (see Figure 4.14).

Holes should be drilled initially in sets of six or more with a 20 mm or larger bit and dust samples collected separately (typically) from the following depth ranges in millimetres: 0–10; 10–25; 25–50; 50–100; 100–150. Each sub-sample used in chemical analysis should weigh at least 5 g and should itself be one of a number of identical sub-samples taken from a **bulk sample** that represents the concrete at a given depth and location. Bulk samples therefore need to weigh 25 g or more and the sub-samples that are not used for immediate analysis should be kept for confirming the analyses, **use by other parties in dispute**, or replacement in case samples are lost. In the shallow ranges, six 20 mm holes are barely enough for reliable analysis but not all of the holes need to be deepened in deeper ranges. In each location the samples will represent only a tiny fraction of the concrete of the structure and the sampling scheme should be designed to give confidence that any instances of chloride contamination will be found. The failure to take enough or sufficiently representative samples is a far greater cause of misleading information on chloride determinations than inaccuracies in the chemical analysis itself.

Tip. Drill dust is easy to collect in a paper pocket stuck just below the drill hole with masking tape. Clean the surface with spare tape before trying to stick to it.

With the right equipment, **chemical analysis by automatic potentiometric titration** or **X-ray fluorescence**, can be set up on a production-line basis that allows large numbers of samples to be analysed quite quickly by relatively few skilled technicians. Although test kits for chemical analysis like **Hach** and **Quantab** designed for quickly testing aggregates in the field can be adapted for use on concrete samples, adapted tests take a long time to perform and are too limited for anything other than initial assessments. Using field methods is unlikely to encourage the proper care of samples and also unlikely to be accurate enough in borderline cases.

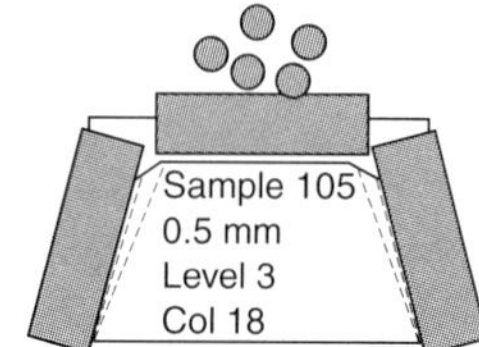

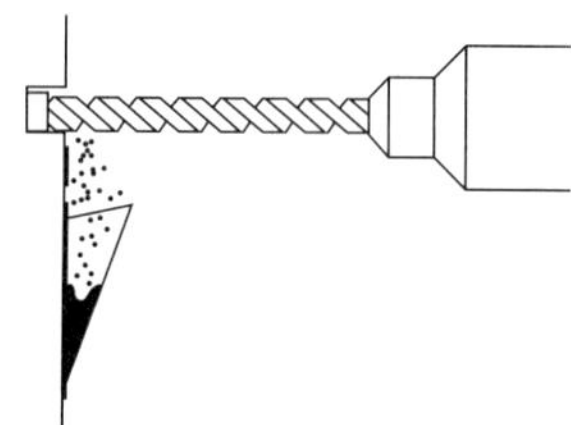

Figure 4.14 Dust samples can be caught in a paper pocket taped to the surface of the concrete.

In very wet or **saturated concrete**, chloride contamination may lead to loss of reinforcement as **soluble iron compounds**, e.g. **'black rust'**. It is not possible to take drill dust samples in these conditions and chloride concentration (if it needs to be known) is best confirmed by breaking out sample areas for inspection.

4.8 Tests on in-situ concrete

4.8.1 Rebound hammer

Cube strength is of almost obsessive interest to designers because it is a simple measure of the potential strength of the concrete, but it is less interesting to investigators who are concerned with actuality rather than potential and with **durability** more than with strength. A rough estimate of the **quality of concrete** can be made non-destructively by testing the surface with a **rebound hammer** (for example **Schmidt hammer**). Quite a lot of readings have to be taken to get useful data and some interpretation is necessary, but this test can be very useful as a means of comparing one area with another and therefore of identifying suspect areas which need closer examination.

Each test area should not be much larger than about 300 mm × 300 mm and at least ten readings should normally be taken in it. Unless there is some valid reason to exclude a reading, all readings in a group are used to calculate the average.

Although rebound-hammer test results are more useful as a means of identifying areas of unusually poor concrete than as an absolute measure of strength, if **cores** subsequently taken from the tested area are used to calibrate the hammer test results, they can be used for indicating strengths. Reference [7] gives useful information on rebound hammers.

4.8.2 Core samples

Testing **core** samples is the only reliable way to measure the **concrete quality** and cores allow many other (and more useful) measurements to be made as well. Cores can yield samples of the reinforcement and they can be used to determine the **depth of carbonation**, **chloride penetration**, the density of the concrete, the **cement content and type**, the effectiveness of compaction, **aggregate grading** and **concrete permeability**. Only a few of these are normally likely to be needed to help with repair **decisions**, but where structural load capacity is in doubt it is important to know the compressive strength (see Figure 4.15).

In **disputes**, cores are invaluable as a means of providing the parties with lasting **evidence** that can be examined and re-examined many times if the need arises.

Cores are quite expensive to cut, so it is worth giving a lot of thought to identifying the best positions to take them from. Non-destructive tests can help to identify locations

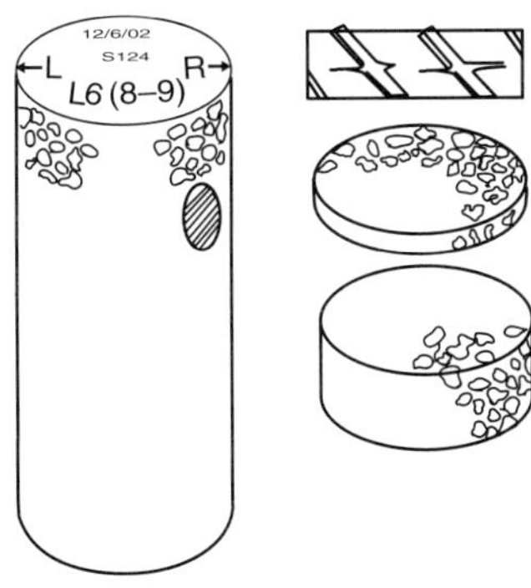

Figure 4.15 Cores are the ideal samples: they allow many tests to be made and provide permanent evidence.

where cores are likely to give a clear picture of the causes of deterioration. There are also obvious considerations like avoiding weakening the structure by unintentionally cutting through reinforcement, or causing damage by severing cast-in services.

The location and orientation of the core should be marked before it is cut and should be confirmed as soon as it has been freed from the bit: it can be difficult to work it out after eager onlookers have turned the core over a few times. **Photographs** and sketches of the core's features made immediately after extraction can be invaluable too, especially in cases where corrosion may take place quickly if previously very wet chloride-contaminated reinforcement is exposed to air.

Core holes can give as much information as the cores themselves, and it is very important to examine the holes and sketch, photograph and note any significant features. Some form of portable lighting and an inspection mirror are usually essential for this. Holes sometimes show the position and orientation of cracks and **delamination** more clearly than the cores (because the investigator sees both sides of the hole at once) and if a core disintegrates during extraction, the hole will still be there as evidence which preserves the relative positions of features [7].

4.9 Exposing reinforcement

It is unusual to assess **structural load capacity** during investigations of deterioration caused by corrosion, but is likely to be necessary if the corrosion damage is incidental to other damage, e.g. **fire** or impact, or in cases where severe loss of section may have been caused by **chloride contamination** in very wet concrete. If structural capacity is to be assessed, it may be necessary to expose and examine several layers of the reinforcement (especially shear reinforcement) in small areas. The only satisfactory way of doing this without disturbing the reinforcement is by **blasting with high-pressure water**. This must be done with great care, and support will be needed if concrete is to be removed in stressed areas.

4.10 Resistivity tests and other measurements

Several other means of measurement and testing are available. They include measurement of electrical **resistivity**, measurement of **surface water absorption**, determination of **elastic modulus** by **ultrasonic pulse velocity**, **radar scanning** and **corrosion current** measurement.

The most commonly used of these are tests to measure resistivity. Earlier sections have shown the importance of resistivity in controlling the rate of corrosion, but unfortunately it is difficult to measure directly at the level of the reinforcement and also difficult to interpret results because they are considerably affected by seasonal changes. Gross differences in resistivity which place the likelihood of significant corrosion into different categories (see Table 2.2) can be identified, but they are likely to be obvious to the experienced investigator simply from the apparent quality of the concrete.

Resistivity at, or near, the surface of the concrete is usually measured by passing an alternating current (to avoid electrolytic polarization effects) of known value through the concrete and measuring the voltage drop. Two- and four-probe instruments are available and there are several methods of ensuring good electrical contact with the concrete. This and the other measurements can be useful if they are made by experienced specialists and they may also be helpful in confirming areas where the less contentious, but more destructive tests should be carried out. However, many experienced investigators agree that the interpretation of their results is difficult. The subject is covered in a helpful way in reference [7].

4.11 The report of the investigation

The main purpose of the **report** is usually to give enough information about the nature and causes of all the different types of **deterioration** to allow **decisions** to be made on what repair methods would be suitable, though it may be too early to specify the quantity of repair work of each type: this tends to be confirmed only once the repairs are under way.

Once the detailed **investigation** has been completed and reported, the **client** often becomes more relaxed about the urgency of taking action. For this reason it is useful if the report records the condition of the structure in sufficient detail for changes in condition to be identified if a further investigation is undertaken at a later date, which may well be several years in the future.

The length and complication of the report will depend very much on the circumstances. For a modest structure, a report which records only a few typical examples of deterioration, rather than a detailed description of each occurrence, may be sufficient and amount to several pages of description, values obtained from measurement and their analysis, drawings, photographs, references, inferences and conclusions.

The report of the investigation may stop short of specifying what repairs are to be carried out, but unless its sole purpose is to provide **evidence** and expert opinion in a dispute, it must provide enough information for the client to compare the technical, **economic and strategic** merits of their various **options**.

A report prepared for a **dispute** will form a key part of the expert **evidence**. Quantities and individual occurrences of damage will have to be recorded and sources of supporting references will have to be cited. All statements of fact will have to be thoroughly checked. On large and complex structures preparing such reports can require the work of a team of experts and the output can run to several volumes.

Checklist of briefs for investigations

Formalities

client's organization, address, phone, etc.

name of person authorized to act for client organization, address, phone, etc. a single point of contact plus a named deputy has advantages.

identity of structure to be investigated.

owner of the structure to be investigated the client may be an agent, leaseholder or tenant.

client's authority to commission the investigation see above.

investigator's organization, address, phone, etc.

name of person authorized to act for investigator's organization, address, phone, etc. a single point of contact plus a named deputy has advantages.

investigator's right of access to structure the investigator may need to show evidence of permission to enter premises, and this may be restricted to defined areas; when investigating public works such as highway bridges it is advisable to inform the local police.

confidentiality of information whether any part of the work is confidential and to whom (in disputes all information is likely to be required for disclosure in the 'discovery' process).

Specially for disputes

organizations involved with the structure, past and present the investigator should be able to confirm that he or she is independent and free to act for the client (essential where disputes may arise).

previous investigations on this structure commissioned by client it is not unknown for clients to commission repeated investigations in the hope of eventually getting results that suit their purpose.

known previous investigations on this structure commissioned by others an investigation may be commissioned to confirm or contradict investigations commissioned by others.

investigator's involvement with known parties in dispute the investigator should be able to confirm that he or she is independent and free to act (essential where disputes may arise).

Scope and definition of tasks

objectives of investigation should be defined, e.g. for initial investigations, information which allows the client to make specific initial decisions; for disputes, evidence and expert opinion; for detailed investigations, information which allows the client to evaluate options.

limitations to what areas will be investigated defining what the investigation will not include; e.g. inspection of only exterior parts that can be reached without access equipment; or inspection of only sample areas showing the greatest deterioration.

limitations to what methods and equipment will be used e.g. visual inspection (a); or (a) + tests on sample areas with a covermeter and phenolphthalein spray (b); or (a) + (b) + surface electrode potential mapping of sample areas plus chloride contamination tests at points of lowest potential.

distribution of reports e.g. number of copies; recipients for copies; confidentiality status; security numbering; electronic distribution (see reports checklist).

Contractual points

date to start it may be appropriate to specify penalties for a late start.

date to finish it may be appropriate to specify penalties for late completion.

intermediate and final reporting dates intermediate reports may be required on certain dates or after specified work has been done (for example, a report summarizing each visit to site).

invoicing and method of payment when the investigator will submit invoices (for example, after each interim report) and when he or she will be paid.

responsibility for contractors if the investigator is to use contractors supplied by others (e.g. the client) the brief should state how their work is to be specified and supervised.

recipients of samples the client may wish to specify named parties who are to be given additional samples.

responsibility for storage of samples a long periods of inaction (sometimes years) can follow the completion of an investigation; the brief should specify the maximum time the investigator is responsible for storing samples and anyone to contact before disposing of them.

making good areas where concrete has been removed, e.g. by chipping or drilling, may need to be left for inspection by others; someone other than the investigator may have the responsibility for making good.

resolution of dispute between client and investigator a mutually acceptable mediator or arbitrator (often an institution) to whom disputes should be referred in the first instance.

law to apply the law to be applied in interpreting the brief as a contract.

Travel, accommodation and equipment.

information to be supplied by the client drawings and documents relating to the design and construction of the structure and any information relating to its condition, previous investigations or allegations.

provision of secure storage for investigator clients are often able to use local knowledge or local contacts to make facilities available to investigators; arranging for them to do this helps to involve clients in the investigation as well as benefiting investigators directly.

provision of office space see above.

provision of workshop space see above.

provision of local equipment and transport where the investigator is working a long way from home, especially if visits are made by air, costs can be saved if the client is able to provide equipment or transport locally.

Working in occupied structures.

occupiers or users of the structure assessment is usually carried out on structures which are in use; the brief should specify any arrangements the investigator should make to minimise inconvenience to users.

investigators working hours working hours on site may be restricted by the need to limit disturbance to users or occupants, or interference with processes or traffic.

occupant's representative on occupied sites the investigator should liaise with anyone designated to represent the occupiers.

nuisance limitation where noise, dust, etc. are an unusually sensitive matter, the brief should record any limitations on operations.

security clearance any security clearance required for investigators or their operatives.

Checklist of useful equipment for investigations

access and support equipment if needed, this is something to hire with an experienced operator.

angle grinder if needed, probably a tool to hire with an experienced operator; stop cutting as soon as you see sparks!.

batteries try to use equipment that uses commonly available throw-away batteries which you can buy at any filling station; rechargeable batteries are a great idea until they run out far from home.

binoculars can save a lot of climbing, but magnification of more than ×8 makes handshake a problem.

bristle brush for initial cleaning of concrete surfaces.

camera(s) with spare batteries a compact with zoom is the easiest to use when working close-up from ladders or cradles; an SLR is best for distance shots; take both and perhaps a tripod.

cintride or similar bits (about 3 mm) and suitable self-tapping screws for making connections with reinforcement; these special bits for drilling 'difficult' steels are a great time saver when making connections to reinforcement.

club hammer a wrist strap is essential when leaning out of windows or working over water, etc.

cold chisel keep it sharp; a wrist strap is essential when leaning out of windows or working over water, etc.

compass for confirming orientation, especially inside buildings.

concrete reinstatement kit dry 3 : 1 sharp sand : cement mortar and water for filling drilled holes and inspection areas, unless someone else is to be responsible for this.

core cutter with water supply and power supplies; if needed, probably a tool to hire with an experienced operator.

cover meter with spare batteries can be hired; try it out before setting out.

crack-width gauge or comparator a simple plastic card with lines of different width printed on it, or a vernier calliper; helps to classify cracks by width.

drawings of the structure it's worth spending time to get them if you can; otherwise draw your own.

folding picnic table can turn an awkward corner into a mini-office.

folding rule or metre stick the best way to mark crack spacings, etc. when working single handed.

half cell kit can be hired; try it out before setting out; spare batteries, 20 m or more of stout 'earthing' flex, terminal clips.

hammer drill with 20 mm or larger masonry bits, chuck-key, cables power supply and spare fuses; a smaller cordless drill with a 10 mm bit can be handy too, but to take proper chloride samples you eventually need the big one anyway.

hard hat even on the tidiest site or occupied building it can save your head when straightening-up after peering at some interesting defect.

inspection lamp helpful even in daylight for taking photographs and inspecting shaded areas.

ladders and steps at least a 2 m ladder is useful for reaching a convenient working height on almost any site, longer ladders on some sites; tie ropes or a lightweight (e.g. Ankalad) stabilizer are essential when working alone.

location map useful for orientation, locating nearby sources of salt contamination, and for the report.

magnifying glass (about ×4) useful for looking at poor concrete and when using a crack comparator.

masking tape useful for marking photograph locations, fixing folding rule to concrete, etc. carry plenty of spare tape for pre-cleaning concrete surfaces to get adhesion!

mirror an inspection mirror for awkward places; a small convex mirror (e.g. a car mirror) can also be useful for reflecting light onto a feature being examined or photographed.

monocular scope a telescope-cum-magnifier which can focus as close as 20 mm and magnify up to ×50; some include measuring scales for crack width measurement.

notebook, pens, paper and clipboard there's never anywhere clean to put anything down; a clipboard can be a portable desk.

paint brush for intermediate cleaning of concrete surfaces before final cleaning with masking tape.

phenolphthalein spray check that it works before setting out; soap or scouring powder are usually alkaline enough.

pin hammer a wrist strap is essential when leaning out of windows or working over water.

plastic bags for samples food bags are ideal, especially those you can write on; seal with masking tape.

plastic cups or some alternative for catching drill dust experienced investigators have their own pet methods; a folded paper scoop is one of the best.

pocket knife or box cutter for sharpening pencils if nothing else; remember to buy one when you land if you are flying.

power supply on occupied sites mains power is usually available if you bring an extension lead.

something to sit on feeling comfortable results in better inspection; a seat should be low enough to allow you to rest your clipboard on your knees.

sponge and water-spray useful for showing up fine cracks.

steel rule to use one-handed for measuring carbonation and cover depths; a vernier calliper is useful too.

steel tapes a 3 m tape is the one you use most but a 15 m tape can be useful too.

strain gauge (e.g. 'demec') studs and adhesive cracks which behave as joints have to be identified; small cross-head stainless steel screws can serve as studs if you use a vernier calliper for measurement.

tape recorder with spare cassettes and spare batteries can supplement but must never substitute for written notes, sketches and drawings.

torch with spare batteries always useful, but could save your life when moving about in a burnt-out building.

towels for wiping hands, kneeling on, etc.

vernier calliper has all sorts of uses, see above.

warm and waterproof clothing (especially UK!) feeling comfortable results in better inspection; a body warmer with lots of pockets is useful; a fisherman's umbrella is also a boon when writing notes and making sketches in the rain.

wax crayons, marker pens and blackboard chalks for writing directly on concrete or on masking tape.

Checklist for reports – suggested headings

confidentiality classification in some organizations 'confidential' requires documents to filed under lock and key; check with the client what term is appropriate. (Note that in disputes only 'privileged' documents can be withheld from the 'discovery' process; if the expert is the subject of a subpoena, even these may come to light.)

circulation of copies list the recipients; add the copy number if the circulation is restricted.

title of report hierarchical titling can help with filing and searching on large projects.

status of report e.g. draft, preliminary, provisional, interim, final.

client

author's name and qualifications

name, address telephone number, etc. of author's organization

date of report

summary a brief list of the objectives, conclusions and recommendations may be sufficient.

instructions the authorization from the client to undertake the work; refer to the brief on large projects.

summary of author's qualifications and relevant experience if the author is likely to become an expert witness in a dispute this section should justify his claim to the appropriate expertise.

scope and objectives of this report if this is one of several reports on the project, the scope and limitations of this particular report should be stated.

documents list previous reports on the structure by the same author or organization, and documents referred to (and 'relied on' in disputes) written by others; state who supplied the documents.

account of visit to site date, weather, time on site, who was present, information given by and co-operation received from anyone present (or absent, e.g. in authorizing inspection of site, providing information, equipment, etc.), witnesses to testing, etc.

account of investigation what structure or part of structure was inspected, what was seen generally, what signs of deterioration, etc. were seen specifically, what evidence was seen of previous tests, etc. by others, what measurements were made, what samples taken and the results.

accidents and damage report any accidents or any unavoidable damage done in the course of investigation, and the remedial action taken.

discussion facts and inferences from inspection and testing; reliance which can be placed on them; significance in relation to the objectives of the investigation.

conclusions list (preferably) of relevant conclusions.

recommendations list (preferably) of options for remedial action with comment on their relative merits if appropriate.

future actions suggested or agreed next steps, or proposals for changes to previously agreed plans.

signature of investigator with date in disputes especially, reports become evidence: they should be signed immediately below the text to discourage any additions or alterations.

copy number of report reports with restricted circulation should have the individual copy number added by the signatory.

Checklist of vulnerable locations

Reinforcement close to the surface

At the corners of beams and columns, where the reinforcement (especially links) can be close to more than one surface.

At intersections between bars forming different sections, e.g. where floor slabs join edge beams, because it can be difficult to bend and fix bars to ensure adequate cover for more than one set.

At places where bars are lapped, e.g. starter bars near the base of columns and walls.

Over mechanical splices which have a larger diameter that the reinforcement.

At joints in formwork where misalignment may result in a step change in the depth of cover.

At movement joints and construction joints, where there is often poor compaction and little end cover.

In slender features where there is little room for cover, e.g. drip grooves in sills and overhangs.

At soffits (especially cantilevered balconies) where the reinforcement may have been bent down by workers walking on it.

Poor compaction and materials

At construction joints where stop-ends were poorly constructed.

Where there are signs of bad workmanship or poor materials, like honeycombing or visibly poor-quality concrete.

Where there are signs of earlier repair or 'making good'.

At the tops of pours (including intermediate pours) where settlement of the solids in wet concrete has caused bleeding and left the reinforcement unprotected.

Where there is cracking from other causes

At thermal, shrinkage and settlement cracks which were not caused by corrosion but have reduced the protection given by cover.

Wetting and drying

In tidal or maritime flood zones.

Where down pipes and overflows leak.

Where nearby plants are watered.

At exposed sills and overhangs.

Chlorides

In highway structures, access balconies, footways, etc. where de-icing salts have been applied deliberately (especially 'halved' joints and other places which hold salty water).

In areas (including unsealed joints) wetted by salty run-off, or where salt has been tracked in by vehicle movements (e.g. parking garages).

At projections, ledges and pockets where wind-blown salt or salty dust and sand can accumulate from the sea or nearby salted structures.

Where concrete has been made with sea-dredged aggregates or sea-water.

Where calcium chloride was used as an accelerator in the concrete (it was virtually banned in 1977).

Where permanent wood-wool formers or formwork were used (calcium chloride is used to pre-treat the wood shavings).

Where decorative finishes have been cleaned with hydrochloric acid.

Where sea-water (or sometimes even 'greywater') has been used to fight fires, flush lavatories, or for fire practice.

Areas washed by the tide or in the splash zone.

In some industrial areas, including farms, food preparation and vehicle washing areas.

In areas which have been sprayed with urine (e.g. lamp-posts, bus stops, or even stairwells).

Figure C.1 Misplaced formwork, or reinforcement, or both resulted in loss of cover in this car park in Düsseldorf, Germany.

Figure C.2 Standing on the reinforcement eliminated some of the bottom cover in this cantilever balcony in Bahrain.

Figure C.3 Walking on reinforcement reduces bottom cover on this site in the Far East.

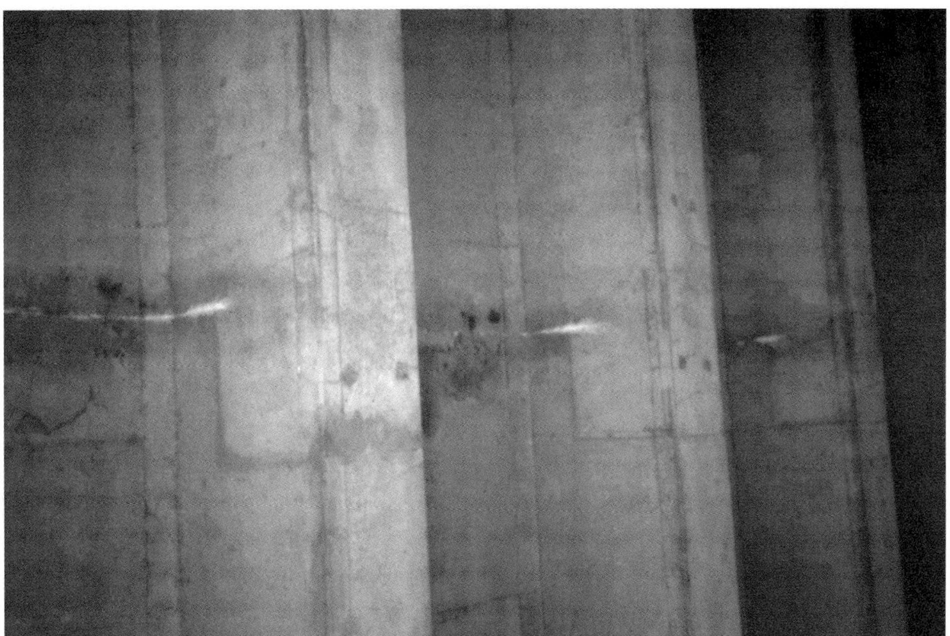

Figure C.4 Motorway in Montreal, Canada: rust staining at construction joint in newly built section; staining probably caused by rusting debris on soffit formwork.

Figure C.5 An extreme example of leaking formwork in a building under construction in Dubai.

Figure C.6 Motorway in Montreal, Canada: damage caused by salt run-off leaking through a poorly sealed expansion joint.

Figure C.7 Corrosion caused by run-off of de-icing salts on low cover elements of a UK motorway

Figure C.8 Credit River Bridge, Ontario, Canada: repair of damage caused by salt run-off.

Figure C.9 Credit River Bridge, Ontario, Canada: repair of damage caused by salt run-off. The inner arches were exposed before the bridge was widened after several years in use.

	Method	Significant chloride contamination by mass of cement, depending			
		>0.4–0.5% in the mix		>0.2% penetrated	
				Over large areas	
		FOR	AGAINST	FOR	AGAINST
1	Add extra cover with cementitious materials to provide a carbon dioxide resistant layer or restore alkalinity		• has no effect		• has no effect
2	Replace affected cover with cementitious materials to restore alkalinity or uncontaminated cover	• simple proven technology • has potential for long-term maintenance-free restoration	• all affected concrete within 20 mm of any reinforcement must be replaced • very difficult to clean behind reinforcement • vulnerable to diffusion from untreated areas • high first cost	• simple proven technology • has potential for long-term maintenance-free restoration	• all affected concrete within 20 mm of any reinforcement must be replaced • very difficult to clean behind reinforcement • vulnerable to diffusion from untreated areas • high first cost
3	Maintain cathodic protection to keep reinforcement cathodic (in relation to an externally applied anode) throughout service life. Usually by impressed current but can be by sacrificial anodes	• has the potential to protect all electrically-connected reinforcement	• the protection must be maintained throughout the service life of the structure • changes profile of treated areas • danger of hydrogen embrittlement in prestressed steel	• has the potential to protect all electrically-connected reinforcement	• the protection must be maintained throughout the service life of the structure • changes profile of treated areas • danger of hydrogen embrittlement in prestressed steel
4	Undertake electrochemical realkalization to increase alkalinity. Requires a short-term application of power through a temporary external anode tank; exposed surfaces may then require coating to prevent loss of alkali by leaching; relatively recent technology		• not the right method for this purpose; although all electrochemical processes listed here can have beneficial side effects as a result of ion migration these may not provide much protection		• not the right method for this purpose; although all electrochemical processes listed here can have beneficial side effects as a result of ion migration these may not provide much protection
5	Undertake electrochemical chloride extraction to reduce chloride ion concentration near reinforcement		• not usually possible to ensure that remaining chloride does not diffuse into treated areas	• all electrically connected reinforcement can be treated	• if chloride has penetrated deeply it may not be possible to ensure that remaining chloride does not diffuse into treated areas

(see Chapter 5 for more information)

on cement type, carbonation, etc.		No significant chloride contamination	
from the environment		Carbonation in affected areas and those at risk	
Locally			
FOR	AGAINST	FOR	AGAINST
	• has no effect	• can provide long-lasting maintenance-free protection against future carbonation	• changes profile of treated areas. • first cost greater than METHOD 6 • effect on existing carbonation is unproven
• simple proven technology • has potential for long-term maintenance-free restoration	• all affected concrete within 20 mm of any reinforcement must be replaced • very difficult to clean behind reinforcement • vulnerable to diffusion from untreated areas • high first cost	• often best value for reinforcement already in carbonated concrete • necessary where cover has been physically damaged • simple proven technology • has potential for long-term maintenance-free restoration	• patchy appearance unless followed by coating to protect areas at risk and to give uniform appearance • cover may need to be increased beyond original profile of member
• has the potential to protect all electrically-connected reinforcement	• the protection must be maintained throughout the service life of the structure • changes profile of treated areas • danger of hydrogen embrittlement in prestressed steel • other methods may be more economical for small areas	• has the potential to protect all electrically-connected reinforcement	• the protection must be maintained throughout the service life of the structure • danger of hydrogen embrittlement in prestressed steel • unlikely to be economical unless incidental protection required for areas which are contaminated with chloride
	• not the right method for this purpose; although all electrochemical processes listed here can have beneficial side effects as a result of ion migration these may not provide much protection	• can restore alkalinity as far as the depth of any electrically connected reinforcement	• may need barrier coating to be maintained to prevent recarbonation • other methods may be more economical, especially for limited areas
• all electrically connected reinforcement can be treated	• unless exposure is changed, a barrier coating, sealing or surface treatment to resist chloride penetration must be maintained throughout service life		• not the right method for this purpose; although all electrochemical processes listed here can have beneficial side effects as a result of ion migration these may not provide much protection

	Method	Significant chloride contamination by mass of cement, depending			
		>0.4–0.5% in the mix		>0.2% penetrated	
				Over large areas	
		FOR	AGAINST	FOR	AGAINST
	requires a short-term application of power through a temporary external anode tank; relatively recent technology				• unless exposure is changed, barrier coating or surface treatment to resist chloride penetration must be maintained throughout service life
6	Maintain coating throughout service life to resist carbon dioxide penetration. Requires monitoring and maintenance; proven technology		• has no effect		• has no effect
7	Maintain moisture resistant coating or surface treatment throughout service life to keep resistivity high. Requires monitoring and maintenance; relatively recent technology; usually also provides protection against future carbonation by METHOD 6 above		• recent research suggests it is not possible to control corrosion currents in this way in concrete already contaminated with chloride		• recent research suggests it is not possible to control corrosion currents in this way in concrete already contaminated with chloride
8	Maintain sheltering (e.g. overcladding) throughout service life to keep resistivity high. Overcladding is a popular application for buildings; widely accepted technology	• not yet known whether it is possible to control corrosion in chloride-contaminated concrete in this way	• not yet known whether it is possible to control corrosion in chloride-contaminated oncrete in this way	• not yet known whether it is possible to control corrosion in chloride-contaminated concrete in this way	• not yet known whether it is possible to control corrosion in chloride-contaminated concrete in this way
9	Replace affected members or rebuild entirely. Potentially foolproof and permanent; need involve no new technology	• foolproof and permanent • can also provide opportunities for upgrading performance and improving appearance	• high first cost	• foolproof and permanent • can also provide opportunities for upgrading performance and improving appearance	• high first cost

(see Chapter 5 for more information) *(contd.)*

on cement type, carbonation, etc.		No significant chloride contamination	
from the environment		Carbonation in affected areas and those at risk	
Locally			
FOR	AGAINST	FOR	AGAINST
• typical applications are salt-water intrusion at construction joints or local failures in waterproof membranes			
	• has no effect	• low first-cost protection where reinforcement is not yet in carbonated concrete • if the coating also resists moisture penetration can control corrosion even where reinforcement is already in carbonated concrete (see below)	• coating must be monitored and maintained throughout service life
	• recent research suggests it is not possible to control corrosion currents in this way in concrete already contaminated with chloride	• effective in concrete which is already carbonated • added protection if the coating also resists carbon dioxide penetration for areas at risk (METHOD 6), as is often the case	• coating must be monitored and maintained throughout service life
• not yet known whether it is possible to control corrosion in chloride-contaminated concrete in this way	• not yet known whether it is possible to control corrosion in chloride-contaminated concrete in this way	• effective in controlling corrosion rate provided cavity conditions are suitable • also provides opportunities for upgrading performance and improving appearance, especially in buildings	• high first cost
• foolproof and permanent • can also provide opportunities for upgrading performance and improving appearance	• high first cost • unlikely to be as economical as other methods	• foolproof and permanent • can also provide opportunities for upgrading performance and improving appearance	• high first cost • unlikely to be as economical as other methods

5. Repair

Repair of **defects** caused by reinforcement corrosion should not be the responsibility of those without appropriate specialist knowledge and experience. This chapter outlines repair processes which are in common use and is intended to help investigators and **clients** to understand what the various repair **options** entail. Anyone involved in the protection and repair of concrete structures, whether with **site-batched materials** or **proprietary materials, products and systems**, will find authoritative guidance in the newly available European Standards for products and systems for the protection and repair of concrete structures [14]. See Chapter 6 for more information.

5.1 Options and strategy

If a structure has **defects** caused by reinforcement corrosion, **clients** have a number of **options** to consider in conjunction with their specialist advisers. Even an initial assessment should be able to tell them a lot about the causes and seriousness of the defects and the prospects for successful repair. A detailed **investigation** should allow them to estimate the viability and cost of the various options.

Corrosion damage which results from **accidental overloading**, from **fire damage**, from **faulty construction** or poor materials, can often be repaired to restore the **service life** fully to that originally intended by the designer. Corrosion damage which results from **chloride contamination** on the other hand is more difficult to repair to a standard which restores long-term **durability** without ongoing expenditure on **monitoring** and **maintenance**. It is important that clients understand whether there is a risk of further deterioration, whether repair may be needed again at some time in the future, and whether they will need to provide for continuing expenditure on maintenance after repair has been carried out.

The risks are difficult to quantify. Although there are now several years of experience in carrying out repairs to concrete structures by engineering methods, there is limited long-term experience of how well these repairs last. The repair options considered in the 'Summary of main protection and repair options' concentrate on those methods which have become established through repeated use, but newer methods are being developed all the time and the client should not reject them provided that there is evidence that they could be appropriate in the particular circumstance. Repair **decisions** and strategies therefore have to be based on educated judgements of uncertainties, and should be made by the client in consultation with expert independent advisers.

5.2 Specifications and contracts

Repair **specifications** have to rely on the results of the detailed **investigation** and testing of the structure for information on what kind of work needs to be done (see Chapter 4). If the investigation was done thoroughly, it should be possible to specify the different types of protection, repair, or replacement in detail, though it is very unlikely that the quantities will be known with much accuracy until repair operations are under way.

If the investigation was less comprehensive, as is often the case if it has been decided not to provide complete **access equipment** until the repair contractor can use it, further investigation and testing will have to be done while the repair proceeds and it may not be possible to specify more than an outline of the type of work that will have to be done.

In any case, the **contract** will have to be a great deal more flexible than most normal construction contracts and expert **supervision** will have to be constantly available make decisions and to control and record the type and amount of work. Provisional items may be unavoidable and remeasurement and 'cost plus' work should be considered acceptable for this type of contract. It is as likely in corrosion repair as it is in human surgery that estimates of the full nature and extent of the problems will change once the patient has been opened up.

The Concrete Repair Association's *Standard Method of Measurement for Concrete Repair* [15] gives an insight into what a repair contract will involve and what uncertainties have to be accepted. This information should be helpful to those choosing between the different repair options.

5.3 Public and client relations

Repairs nearly always have to be carried out on structures that are in use. In the case of buildings, the people who occupy them will be trying to work, learn, sleep or recover from sickness while the work is going on.

A little trouble taken to consult and inform the **occupiers** about what is going to happen to their environment can work wonders and prevent the endless aggravation and interference that can make working on occupied sites an expensive nightmare (see Figure 5.1).

Figure 5.1 Public relations on this British housing repair site were exemplary: the returning family show the contractor what they bought on their shopping trip.

In plant which is operated continuously, repairs can sometimes be divided into operations which can be done while the plant continues to operate and those which will have to wait for a **shut-down**. The same applies to airfield, highway or railway sites where working at night while traffic is light may be an additional option.

Traffic management on highways is a well-understood operation, but there are sites with less familiar requirements. Where there are special security requirements or exceptional hazards, for example in oil refineries, the operators' security and **safety** requirements may require operations of certain types or in certain areas to be restricted to times when the operator can provide special services. Choosing methods that reduce risk or take advantage of any possibilities to separate restricted operations from those which can be done at any time or place should be part of planning the work. A **repair specification** must take these operational matters into account and include enough detail to allow contractors to price them.

5.4 Choice of protection and repair methods

For any method of protection and repair to be effective in the long term, it must either restore **passivity** by removing the causes of corrosion (for example by removing **chloride contamination**), or prevent corrosion through **cathodic protection** or **anodic control**, or be based on at least one of the two proven principles of controlling **corrosion currents**, that is, **cathodic control**, or **resistive control**. These fundamental principles of protection and repair, and the methods which make use of them, are central to reference [16] (see Chapter 6). All methods have features that give them advantages and disadvantages in specific circumstances and methods which are in common use are summarized in the 'Summary of main protection and repair options' in this guide. Additional methods which comply with the fundamental principles of protection and repair are included in this reference, and in such an intensively researched field new methods are frequently being proposed and developed. Evidence of satisfactory performance should be sought by any **client** who considers using a new method other than on a trial basis.

Non-technical as well as technical matters affect the choice of options. For example, if it is thought that it might be possible to remove all the chloride-contaminated concrete which could endanger reinforcement (a fact that might not be known for certain anyway until concrete removal and testing are under way), the relative lack of disturbance that **electrochemical protection and repair methods** cause to **occupiers** of buildings could tip the balance in favour of the alternatives of **cathodic protection** or **electrochemical chloride extraction**, say in hospitals or hotels. Conversely, a once-and-for-all repair method could have decisive advantages over one which requires ongoing power supplies and **monitoring** in the case of a structure in a remote location where there is no easy access or reliable power supply. Another case might be a building which has a low value or attracts a low rental only because of an outdated or unappealing **appearance** and perhaps poor insulation. Here the more expensive option of **over-cladding** could be a better long-term investment than cheaper 'traditional' **patch repair**.

This guide is concerned only with satisfactory long-term protection and repair made by sound engineering methods. It is accepted that there are cases where the client will choose the option of palliative treatment that has regularly to be repeated. Such treatments might include patching without removing chloride contamination or replacing **carbonated concrete**, or painting exposed reinforcement simply to prevent it from **rusting** away, but without providing anodic control. These treatments can have their place if they are chosen with full knowledge of their drawbacks, but they are not protection or repair in the sense that they can restore the structure to an acceptable condition for the remainder of its **service life**. They are outside the scope of this guide.

5.5 Repairing cracks: delamination and spalling

5.5.1 Plastic shrinkage cracks

Plastic shrinkage cracks are described in Section 3.2.1. Plastic shrinkage cracks in concrete surfaces exposed to **salt** should be examined by **sampling** at points near any reinforcement to see if the cracks have allowed salt to penetrate the concrete. If results show that there is a danger of significant **chloride contamination** within 20 mm of any reinforcement under the cracks, the contaminated concrete should be cut out and replaced at these points unless **electrochemical protection and repair methods** are to be applied. (If cracks of this type occur during construction where there is salt exposure, they should be sealed as quickly as possible to prevent salt from getting in.)

5.5.2 Plastic settlement cracks

Plastic settlement cracks occur as a result of aggregates in over-workable concrete settling in the formwork after the concrete has been compacted, and cracking the concrete where it hangs over the reinforcement as it stiffens. Because the cracks run parallel with the reinforcement they can be a serious cause of corrosion. Plastic settlement cracks are described in Section 3.2.1.

If corrosion has disrupted the concrete at these cracks, all contaminated concrete within 20 mm of the reinforcement should be cut out and replaced unless an **electrochemical method of protection** and repair is to be used. (If these cracks are found in new concrete, it may be possible to inject some parts of them underneath the reinforcement with modified cementitious grout, though it would be difficult to be certain of filling all the voids in this way.)

5.5.3 Transverse cracks

Transverse cracks are not usually caused by corrosion and unless they are wide they seldom responsible for corrosion: they are described in Section 3.2.1. However, the positions of transverse cracks can be influenced by adjacent bars in a lower layer of reinforcement acting as **crack inducers** and in this case the *lower* bars will be affected by longitudinal cracking (see Section 5.5.4 below).

In new construction in **salty environments**, if transverse cracks wider than 0.3 mm are found before salt has entered, they can be sealed by resin injection unless they are **active cracks**, in which case they must be converted to joints. If salt has entered, injection will not prevent corrosion and the concrete cover will have to be cut out and replaced. **Crack injection** is a well-developed procedure and is undertaken by many specialist contractors. It can be successful in preventing further damage in uncontaminated concrete, but it will not stop corrosion from getting worse if it has already started in chloride-contaminated concrete.

5.5.4 Longitudinal cracks

Section 3.2.1 discusses the causes and dangers of longitudinal cracks.

Longitudinal **cracks** caused by corrosion are common in structures affected by **carbonation** or by **chloride contamination** that reaches the reinforcement. Such cracks cannot be treated without removing and replacing the concrete cover, at least locally, unless an **electrochemical method of protection and repair** is to be used.

5.5.5 Live, or active, cracks

Active cracks are wide transverse cracks that continue to move with changes in load or temperature. They are not controlled by the reinforcement. See Section 4.4.

This type of crack cannot be treated by **crack injection** or grouted successfully because stress concentrated at this point will re-crack it. Where there is a danger that **carbonation** or chloride contamination will enter an active crack and cause corrosion, it must be widened and converted into a sealed **joint**. If a crack is already contaminated in a **salty environment** and in a position likely to cause corrosion, the concrete should be replaced on both sides and incorporate a sealed joint.

5.6 Removing concrete cover

Where concrete has significant **chloride** contamination or where **carbonation** has reached the level of the reinforcement (or is very close to it), the concrete cover should be removed and replaced with new **cementitious** concrete or mortar to ensure that the steel is in a protective environment unless **electrochemical protection and repair methods are** going to be applied.

Unless the area or depth to be removed is small, the **structural** adequacy of the member that will remain after cover has been removed should be considered. In some cases **load-bearing** members can be tackled safely by planning the work so that cover is removed and replaced progressively a small area at a time, but in others alternative temporary **support** will be needed.

Each case will have to be examined individually against the requirements of current codes (if they are applicable) and the apparent assumptions of the original design.

The usual way of removing concrete cover in small areas is with **power hammers**. Sharp tools minimize damage to the concrete that is left but even careful mechanical removal may leave **micro-cracking** which can reduce the **bond** of repair materials. If **pull-off tests** show that the **surface tensile strength** is insufficient for successful repair, the micro-cracked layer must be removed by water or grit **blasting** (see Chapter 6). Power hammers can also easily damage or even cut through reinforcement, so great care must be taken when they are used. Both methods of removal have been shown to allow full composite action to be achieved in load-bearing repaired areas [17].

A quicker way of removing concrete, even in small areas, is by **blasting with high-pressure water**. This method removes concrete without causing micro-cracking and leaves an uneven surface which is excellent for bonding (see Figure 5.2). Water pressures from 50 to 110 MPa are used for this work and require precautions to be taken against personal injury or damage to property that could be caused by the jet or

Figure 5.2 Salt-contaminated concrete removed by water blasting from this British tunnel roadway soffit.

by flying debris. Although high-pressure water blasting is far from peaceful, it is much less intrusive than percussive methods because it does not cause vibration to be transmitted through the structure. This may be a factor when considering alternative options for occupied buildings.

Dry methods of concrete removal generate dust that can contain enough **unhydrated cement** to set quite firmly on the substrate if it becomes damp. This dust must either be removed immediately or be removed before repair materials are applied, preferably by water blasting.

5.6.1 Treatment of edges of removed areas

Unless the whole of the cover is being removed, **patch edges** against sound concrete should be cut at an angle between 90° and 135° with the surface to prevent stresses caused by shrinkage and expansion resulting in spalling.

Patch edges are usually cut with a masonry saw or **angle-grinder**, and should be at least 10–15 mm deep if the **depth of cover** allows. If the reinforcement is shallower than this, as it sometimes will be when **deterioration** caused by **carbonation** damage is being repaired, the depth must be reduced and great care must be taken not to damage the reinforcement. Bright **sparks from burning steel** particles are a sign that steel is being cut. Sawing usually leaves the edge too smooth to **bond** with the repair concrete, so the sawn face should be roughened: **blasting with grit** or **water** have been found equally effective [17].

Blasting with a fine high-pressure water jet is an alternative tool that will cut a reasonably clean perpendicular edge. It has the advantage that it does not cut through steel and the face of the cut does not need to be roughened. A water jet can also easily be used to follow a curved or irregular line.

5.6.2 Depth of removal: carbonated concrete

Where the depth of cover which has not been **carbonated** is insufficient to give protection throughout the **service life** but there is no **chloride contamination**, concrete may have to be removed to make room for an effective depth of uncarbonated concrete over all of the affected reinforcement unless **electrochemical protection and repair methods** or **over-cladding** are to be used (see Sections 5.11–5.14).

The depth of cover needed will depend on the **carbonation coefficients** of the remaining and added cover, and whether or not a carbon dioxide-resistant coating will be **maintained**. If the available depth is insufficient, the profile of the member will be changed by the repair.

The detailed **investigation** (see Chapter 4) should have included enough information to indicate which of these matters need to be taken into account when repair **options** are considered. If removal and replacement of cover is the method chosen, the locations where cover is to be removed and the depth of removal are usually decided on site as the work proceeds. Frequent measurements of **carbonation** and the depth of **cover** need to be carried out, especially during the learning period early in a **contract**.

Even if the profile remains unchanged, **patch repair** leaves the surface looking untidy. It is very difficult to achieve a uniform appearance unless all the visible concrete then receives a uniform treatment after the low-cover and carbonated areas have been repaired. A carbon dioxide-resistant coating can both help to protect uncarbonated cover against carbonation in the future and improve the appearance of the structure (see Section 5.10), but where **appearance** is especially important or where changes in appearance are prohibited (for example on heritage buildings), protective **coating** may need to be followed by a special decorative coating to provide an acceptable finish.

5.6.3 Depth of removal: chloride-contaminated concrete

If the concrete is contaminated only with **chloride penetration** from the surface, complete repair may be possible provided the chloride-contaminated concrete (or the chloride alone in the case of **electrochemical chloride extraction**) can be removed entirely where it is within 20 mm of any reinforcement [14] (see Chapter 6 for more information). This is likely to be practicable only where the contamination is local, for example chloride penetration at leaking **joints** or from faults in **waterproof membranes**.

If chloride contamination has entered the concrete over a substantial area, neither removal of contaminated concrete nor extraction of chloride by electrochemical methods is likely to be economic. If the contamination comes from **chlorides in the mix**, sufficient removal is seldom possible. In these cases **cathodic protection** or the replacement of members are likely to be the only satisfactory options.

Where it has been decided that replacement of locally **chloride-contaminated concrete** will be the chosen repair method, **chemical analysis** of **samples** (for example by **dust drilling**) will be necessary as the repair proceeds to ensure that no significant contamination remains within 20 mm of any reinforcement.

Considerations of appearance are much the same as those in Section 5.6.2 above, though there may be no need to change the profile of the member (see Figure 5.3).

5.7 Reinforcement

5.7.1 Cleaning reinforcement

The reinforcement that is exposed when the concrete cover has been removed may need to be cleaned to prevent it from corroding again. If there is no **chloride contamination**, **blasting with grit** or **water** to remove any **carbonated** mortar or concrete, loose rust, or anything that could interfere with **bond** or the complete coating of the reinforcement with new cement paste from the repair material will be sufficient.

Figure 5.3 Adding extra cover produced this interesting profile at a motorway bridge in Montreal, Canada.

If the corrosion was caused by chloride contamination, thorough cleaning with low-pressure water **blasting** right round the circumference and at least 50 mm at each end beyond the affected length is essential unless **electrochemical methods** are to be used. The action of chlorides in destroying **passivity** is one in which chlorides are not consumed but remain active, and a sufficiently high local concentration of chlorides on the reinforcement can continue to cause corrosion even when the contaminated concrete has been replaced. Chlorides can be removed from lightly **pitted** steel by thorough wet grit blasting or by dry grit blasting to the **Swedish SA $21/2\frac{1}{2}$** standard followed by water blasting, but the difficulty of cleaning behind exposed reinforcement makes it necessary first to remove concrete to a considerable distance all round the reinforcement, usually far more than the 20 mm minimum all round distance required to isolate chloride-contaminated concrete. This makes this method of repair suitable in only a very limited number of circumstances [14]. See Chapter 6 for more information.

If reinforcement is heavily pitted it is doubtful whether there is any way of removing chlorides from the bottom of pits and it is best to replace it.

5.7.2 Replacing reinforcement

Deeply **pitted** reinforcement which cannot be cleaned or which has lost significant section should be replaced with new reinforcement complying with the requirements of Eurocode 2 [18].

If the remaining reinforcement is suitable, **reinforcement replacement** by **lapping, welding** or **with mechanical couplers** are acceptable alternatives. Reinforcement that is lapped must have at least the same lap length for **bond** that the original design required. This can lead to the need to remove much more concrete than is needed for other aspects of repair, just to get a long enough lap length.

5.8 Repair materials

5.8.1 The properties of replacement materials

Replacing concrete with a different type of material, or even with different concrete, will affect the behaviour of the member under loading and temperature changes. Material properties that must be considered include **elastic modulus**, coefficient of thermal expansion, shrinkage and creep, as well as **bond** with the **substrate**. Generally, it is best to use replacement materials that have properties like those of the materials they replace. For composite action to be achieved in **load-bearing** situations recent research has found that the elastic modulus should lie within $\pm 10\,\mathrm{kN/mm^2}$ of that of the substrate concrete [17].

Resin-bound mortars have a higher coefficient of thermal expansion but a lower elastic modulus than concrete, even if they are heavily filled, and their use must be considered with care even in **non-load-bearing** applications.

Whenever possible, concrete should be replaced with **cementitious** concrete or mortar as only these materials restore **passivity** to the reinforcement.

5.8.2 Proprietary products and systems

Proprietary products and systems are often more convenient to use than **site-batched materials** when small quantities are needed for repair work and sometimes they are essential. Specifiers must be knowledgeable about the materials that are used and must ensure that the products and systems are suitable for the conditions in which they will be used.

Although **guarantees** are usually offered by the suppliers of products and systems, workmanship is so critical to the success of repairs that guarantees applying to products and systems can turn out to have little value if repairs fail, whether or not **warranties** have been complied with. The value of any guarantees which stipulate warranties which cannot be achieved in practice (e.g. 'all **rust** and contamination must be removed from reinforcement') must be seriously questioned. Chapter 6 covers testing and outlines the specifications that are available or are in preparation.

5.8.3 Trowel-applied and hand-applied cement-based patching mortar

Small areas and sections less than about 100 mm thick are usually repaired with hand- or trowel-applied **cementitious** mortar with or without **polymer modifiers**. Many **proprietary products** are available, some of them based on lightweight aggregates and specially modified to allow thicknesses of more than 40 mm to be applied in one operation without the material sagging on soffits or vertical faces.

The **substrate** should be damp enough to prevent water from being lost from the repair mortar, but free from surface water. The mortar should be worked well into the prepared substrate to ensure that there is no entrapped air and that the reinforcement is coated on all sides with cement paste from the mortar.

Repair mortars are usually applied with a **gloved hand** so that the mortar can be packed tightly into irregular voids and around the reinforcement. Punning the mortar into place with small pieces of wood can help to ensure that it is thoroughly compacted behind the reinforcement. If it is necessary to build up the thickness by applying the mortar in layers, this should be done wet on wet but if any intermediate layer has been allowed to dry out or harden, the surface should be prepared in the same way as substrate concrete before application is resumed.

5.8.4 Recasting with concrete

Recasting with concrete is an excellent way of replacing larger areas of concrete that have been removed during repairs, but it does need suitable **formwork** and placing. It is seldom practicable to cast sections where the minimum thickness is less than about 40 mm.

Freshly exposed **substrates** usually do not need any further cleaning other than being hosed off with clean water to remove any dust that has settled, and then being allowed to become dry on the surface. Substrates which have been left for more than a few hours after exposure should be cleaned by grit or water **blasting** to remove any hydrated cement dust (see Section 5.6).

To give a good **bond**, the substrate must be clean, sound and damp enough at depth not to suck water from the repair concrete, though it should not be running with water on the surface, and the formwork must be drained of standing water. If the surface is accessible just before the new concrete is **poured**, it can be primed with a rich **cementitious** grout brushed well and thinly into the surface, but for recasting, formwork usually has to be fixed too far in advance to allow grout to remain until the new concrete is poured.

Some **proprietary repair systems** recommend the use of special **bonding coats** some of which contain **polymer modifiers**, but there is little evidence that they generally improve bond for cast repairs. Because polymer-modified cementitious grouts dry out more quickly than plain cementitious grout, it is even more difficult to pour replacement concrete before the bonding coat has dried. A grout or a bonding coat that is allowed to dry before the concrete can be cast, will do more harm than good.

5.8.5 Design of recasting mixes and flowing concrete

Mixes for repair by **recasting** can be designed for compaction by vibration or as **flowing concrete** for self-compaction under gravity. In either case mixes must be designed with the same care that is needed for any cover concrete and there are some additional special requirements. It is inevitable that there will be differences in properties between the **parent** and **repair concretes**, but large differences in **elastic modulus**, moisture movement or coefficient of **thermal expansion** will impose unnecessary stresses.

Generally, it is best to use replacement materials that have properties like those of the materials they replace. Where practicable, the **aggregates** in replacement concrete should be similar in properties to those of the parent concrete but the maximum aggregate will usually have to be smaller. It should be at least 5 mm less than the smallest dimension of the section to be recast (often the void between the reinforcement and the parent concrete). For composite action to be achieved in **load-bearing** situations it is recommended that the elastic modulus lies within $\pm 10\,\mathrm{kN/mm^2}$ of that of the substrate concrete (see Section 5.8.1).

Mixes for repair should comply with at least all the relevant requirements of BSEN 206 [19] but minimum requirements for **cement content** and water : cement ratio may need to be exceeded a little. The Highways Agency specifications for repair concrete for compaction by vibration or flowing concrete are considerably higher than BSEN 206 and specify a minimum cement content of 400 $\mathrm{kg/m^3}$ and a maximum water : cement ratio of 0.4, though the 'cement' in this case is 35% OPC and 65% ground-granulated blastfurnace slag [20].

For awkward **pours**, ultra-workable flowing concrete which compacts itself under gravity can be fed into the void from the lowest point of the **formwork** to prevent air being trapped. If the concrete is to flow into place under a gravity head of a metre or less, as will often be the requirement, plasticizer or **superplasticizer** will be needed to produce the required workability and the concrete will have to be placed quickly, usually less than 30 min.

The fines fraction of the mix must be designed to keep **bleeding** to the practicable minimum, and a shrinkage-compensating **admixture** will be helpful in some situations, for example when concreting soffits.

Additives like ground blastfurnace slag, PFA and silica fume can be used to reduce permeability in the same way as in new construction, and many specifications insist on this. Several **proprietary products and systems** are now available, including shrinkage-compensated flowing concrete.

References [20] and [21] give background information on investigation and repair.

5.8.6 Sprayed concrete

Large areas of concrete up to 70 mm thick are most easily replaced by **sprayed mortar or concrete**. The dry spray process (i.e. where water is applied separately but at the same time as the dry materials) is fast and allows good control of where the concrete is placed. It has the disadvantage that the control of the mix and the density depend entirely on the skill of the nozzle-man and it is difficult to produce consistent void-free highly impermeable cover with it.

In the wet-spray process, fully mixed wet materials are fed to the nozzle pre-mixed and jetted onto the surface with a compressed air stream. **Polymer modifiers** are usually used in this process.

In either process, application should comply with relevant standards. It is important that the integrity of the layers is not jeopardized by striking off, trowelling or any other finishing operation on the final surface. For this reason sprayed concrete may not result in an acceptable appearance for all applications.

5.9 Curing

Replacement **cementitious** concrete and mortar usually has a higher **cement content** than material used for new construction and therefore a greater exotherm during the early hardening period. When unmodified cementitious repair materials are used wet curing is always preferable and especially desirable when it is important to avoid early shrinkage cracking. Wet curing also reduces the temperature rise during hardening and therefore reduces **thermal contraction cracking** on subsequent cooling. Wet curing of cementitious repairs should exceed the minimum requirements of EN 206 [19] and it is recommended that it is continued for 8 days from the time of applying the concrete or mortar. Manual curing is impractical for such a period and almost certain to be neglected: absorbent material kept wet by perforated hoses and tightly covered with transparent plastic sheeting (so that it can be seen whether the material is wet) is a satisfactory method.

Where wet curing is not practicable, plastic sheeting well sealed to the surface at the edges or a spray-applied **curing membrane** are alternatives, but they may not prevent early cracking.

Some **polymer-modified** cementitious mortars need a curing regime which includes some initial loss of water to allow the polymer to coalesce. These are almost always **proprietary products** and the supplier's specification should be followed.

5.10 Coatings and surface treatments on concrete

Coatings and surface treatments on concrete have several applications in protection and repair. Used alone they can increase resistance to **carbon dioxide penetration** or **chloride penetration** beyond the values which can be achieved with concrete, and they can reduce moisture content to values at which reinforcement in carbonated concrete (though not chloride-contaminated concrete) is protected for a considerable time by high **resistivity**. A report of recent research at the University of Aston gives useful information on the use of coatings and surface treatments as the primary method of protection [22]. Used in combination with other methods, surface coatings can reduce the re-penetration of chlorides or carbon dioxide into repaired areas as well as improving **appearance**. Designers and specifiers are sometimes reluctant to rely on coatings and surface treatments because they consider their effect impermanent, but if properly applied, **monitored**, **maintained** and periodically renewed, these treatments can contribute to acceptable long-term protection and repair methods. Alternatively, coatings can be used as temporary protection to postpone cracking and spalling while other repair methods are considered.

Film-forming 'breathing' surface coatings (barrier **coatings**) reduce the penetration of both water (and therefore also **salt** in solution) and carbon dioxide. They have some ability to allow water vapour to escape from concrete but because they also delay the drying out of concrete, they should be applied only to concrete when it is dry. The thicker elastomeric coatings are able to bridge cracks to some extent, but existing wide cracks and **active cracks** need to be treated first. Decorative coatings, even with a matt finish, show up **surface blemishes** that would not be noticed on plain concrete and may even pinhole on some rough surfaces. They can give a disappointing appearance unless the surface is carefully prepared. **Smoothing** or 'fairing' **coats** can be applied first to reduce surface blemishes.

Hydrophobic impregnation treatments (e.g. **siloxane**-based treatments) leave the surface pores in concrete open but affect the surface tension so that the penetration of water (and salt solution) is inhibited. They should be applied to dry concrete which they then encourage to dry out further, but apart from leaving concrete looking dry, have little effect on appearance. They have little effect on carbon dioxide penetration.

Protecting concrete with coatings and surface treatments costs less than repairing it by other methods, but it is not the cheap option that it may seem. The surface must be properly prepared, application must be done under suitable conditions, and **products** specifically formulated for application to concrete are relatively expensive: it is essential that they applied in accordance with the suppliers' instructions. Dry **blasting with grit** is an effective way of preparing the surface to receive coatings, though it does open up near-surface 'blowholes' and increases the need for fairing coats. Reference [23] gives useful information on the choice and application of coatings.

Studies of the performance of coatings are relatively recent and there is little long-term evidence to go on. Such evidence as there is suggests that penetrating coatings which react with the concrete and coat the pores with a water-repellent surface (silanes and siloxanes) are likely to have a longer life than other types of coating; they can also act as effective primers for decorative, carbon dioxide-resistant coatings and breathing coatings used to reduce moisture content and control corrosion by increasing concrete resistivity.

Where coatings and surface treatments are relied on as the primary method of protection, monitoring their performance and renewing them when necessary are essential. Ease of access to the affected parts of the structure is therefore an important factor when considering this repair **option**.

5.11 Cathodic protection

Reinforcement which has lost the protection of the uncontaminated **alkaline** concrete that normally surrounds it corrodes through **electrolytic action** unless the **resistivity** of the concrete is high or there is insufficient **oxygen**. Some parts of the reinforcement become **cathodic**, others **anodic**, and reinforcement is lost from the anodic areas as a result of the **corrosion current** which flows between them (see Section 1.2.1). Cathodic protection is the process of applying external electrical power to oppose this action and prevent corrosion by keeping reinforcement cathodic (in relation to an externally applied anode overlay) throughout the **service life**.

Where a structure is substantially undamaged but at risk from corrosion because of extensive **chloride contamination**, or where damaged parts have been repaired but it is not practicable to remove all the chloride-contaminated concrete which might affect the reinforcement, cathodic protection has the potential to prevent further parts of the reinforcement from starting to corrode. It is probably the only method of protection that can do this and in some cases it may also be cheaper than traditional repair methods which need concrete to be removed.

Cathodic prevention is the term for the process of installing a cathodic protection system as a precaution in a new structure. The system is either left unpowered until needed or powered at a very low level. It is used where long-term durability is critical and environmental conditions are potentially aggressive.

The external power needed for cathodic protection can be applied in two ways: by **sacrificial protection** or by **impressed current**. In sacrificial protection power is generated by galvanic action in which an anode made from a metal or alloy more active than iron in the **electrochemical series** is electrically connected to the reinforcement and physically attached to the concrete. The anode is 'sacrificed' and gradually dissolves in any moisture surrounding it and the concrete. Sacrificial protection has been widely used more than a hundred years in other applications where steel is wet, for example ships' hulls, piles for piers, outboard motors and even fence wire, but in parts of structures which are neither immersed in water nor in moist ground, the high resistivity of the concrete surface restricts the sacrificial current to a value too small to give useful protection. For structures exposed in the atmosphere, impressed current protection is the practical option and it has been widely and successfully used since the mid-1970s (see Figure 5.4). Installation is also relatively unintrusive and therefore suitable for **occupied sites**.

The impressed current is usually supplied by a transformer-rectifier system operating from a local mains power supply and delivering direct current at up to about 24 volts and 10–50 amps depending on the area to be protected. The system has to be controlled to deliver the appropriate current to different parts of the reinforcement and embedded **half-cells** (usually silver/silver chloride) are used to **monitor** this and control the system automatically. The protective **current density** at the reinforcement is typically 10–20 mA/m^2 of steel surface and the power requirement of the system, which varies with the density of the reinforcement, generally ranges from about 20 to 200 W/m^2 of concrete surface (see Figure 5.5).

Figure 5.4 Burlington Skyway, Ontario, Canada: site of pioneering work on cathodic protection for bridge sub-structures.

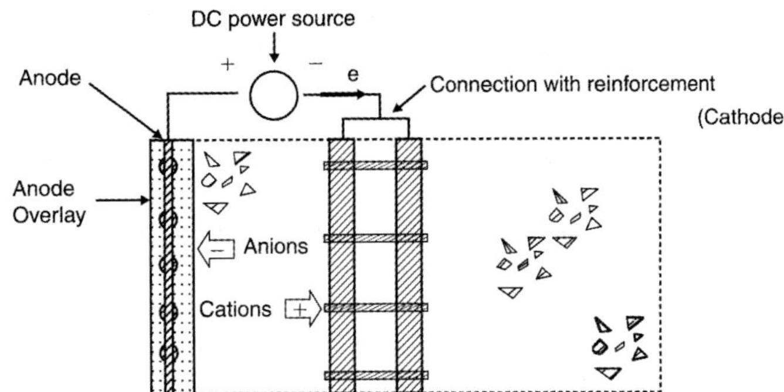

Figure 5.5 Cathodic protection keeps all the reinforcement cathodic and causes slow beneficial migration of ions.

The negative of the power supply (it is *supplying electrons* to the reinforcement to ensure that it is *cathodic*, see Section 1.2.1)) is connected directly to the reinforcement at a number of different locations depending on how well the bars are electrically interconnected and the positive of the power supply is connected to an external **anode** overlay in electrical and physical contact with the concrete surface. The whole system is an electrolytic **cell** *consuming* power from an external source: the anode is therefore positive and the cathode negative (see Section 1.2.1).

The anode overlay is a deliberate target for corrosion and must be designed either to have a long life in the very corrosive conditions that the process will create, or be cheap and easy to replace. Titanium mesh is the basis of the most widely used long-lasting anode overlays and **conductive paints** containing carbon, the basis of probably most replaceable anode overlays, though sprayed zinc is widely used in USA. Preparing the surface to receive the anode overlay and applying the overlay, usually with **sprayed concrete**, are the most intrusive operations in cathodic protection which in general is much less intrusive than other methods of protection and repair apart from surface **coating** (which is in any case not effective for chloride-contaminated concrete). Obviously the overlay, which can be up to 30 mm thick, will alter the appearance of the structure.

Other anode systems include embedded electrodes, electrodes clamped or bolted to the surface, conductive ribbon installed in sawn slots and conductive concrete and asphalt. They all have applications in special situations but have not been as widely used on structures as titanium mesh-based anodes. As with other electrochemical repair methods, which are discussed below, cathodic protection is still currently the subject of intensive research and development.

The protective current must be carefully controlled to ensure that there is a large enough negative potential to ensure that all of the reinforcement is cathodic, but it must not be so large that it will cause **hydrogen evolution** by electrolysis of the **pore fluid**, or any undue risk of undesirable **side-effects** as a result of the migration of **ions** under the influence of the applied potentials. Undesirable effects might include **hydrogen-embrittlement** of high-yield steels or alkali reaction with susceptible **aggregates**, or **reinforcement bond loss**. For this reason **electrochemical protection and repair methods** are not suitable where concrete is prestressed.

At present it is generally accepted that the protection system must be kept running throughout the **service life** of the structure, but any process which passes an electric current through concrete causes the ions in the pore fluid to migrate slowly. There is evidence that **ion** migration caused by cathodic protection that has been discontinued after running for several years may have a longer-term benefit by reducing the chloride ion concentration and increasing the alkalinity at the surface of the reinforcement. Broomfield's book [24] includes much useful information on cathodic protection, and indeed on most aspects of the electrochemistry of reinforcement corrosion.

5.12 Electrochemical chloride extraction

Just as it is possible to apply electrical power externally to ensure that all reinforcement is **cathodic** and unable to corrode, with a rather more intensive but short-term application of power it is also possible to move **ions** through the concrete intentionally. In this way **chloride** ions (Cl^-) in the cover region can be brought to the surface where they can be collected in a suitable **electrolyte** (for example **sodium borate solution**) and removed. (See Section 5.13 for a converse process in which **alkaline** ions are generated in the cover concrete to restore its alkalinity.)

As with **cathodic protection**, an external **anode** is applied to the surface of the concrete. It provides a positive electrical field to attract chloride ions (they are anions, i.e. carry a negative charge) and the negative field at the reinforcement repels them. Current densities for chloride extraction are far greater than for cathodic protection, typically 1 A/m^2 of the reinforcement, at an **application voltage** of 20–100 V. The total charge, which is applied over a period of 2–10 weeks, is about 2000 Ah/m^2 of concrete surface, similar to the total **charge** which would be used in cathodic protection over a period of about 20 years (see Figures 5.6 and 5.7).

Anode systems for electrochemical chloride extraction are temporary and are removed after the process has been completed. The electrolyte can be applied in a simple tank constructed as a frame clamped to the concrete surface, or in spray-applied absorbent material (see Figure 5.8). The need for only simple methods for applying the anode system make this method even less intrusive than cathodic protection. Although they need to have a life of only a few weeks, anodes have to be made from durable materials like titanium mesh to survive the process.

Unless it is possible to get at both sides of a unit (sometimes possible for example with balconies, culverts, etc.) the effect of the electric field cannot be felt beyond the deepest layer of connected reinforcement, and even between this and the anode it has not been found possible to remove all chloride contamination by this method. There is therefore a danger that chloride removed from the immediate vicinity of the reinforcement will eventually be replenished by diffusion from the surrounding concrete. This need not be a problem if chloride penetration was only local and such cases as

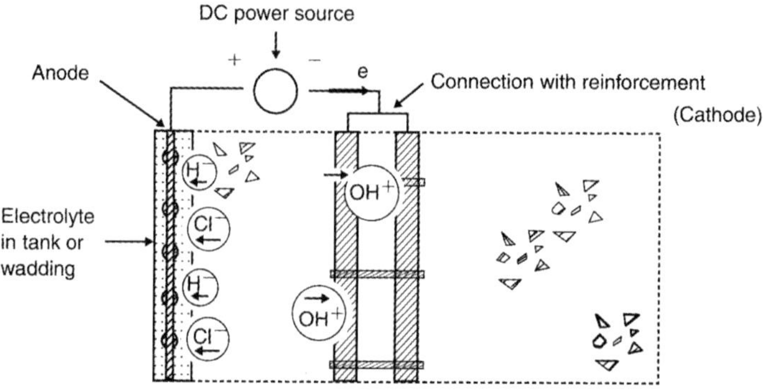

Figure 5.6 Electrochemical chloride extraction causes chloride ions to migrate quickly to the temporary anode.

Figure 5.7 Controlled power supplies to soffit tanks being used in electrochemical chloride extraction.

Figure 5.8 Soffit tanks being used in electrochemical chloride extraction on a British housing site.

penetration at **joints**, **cracks** or faults in **waterproof membranes** are probably the best applications for the method.

Possible undesirable **side-effects** are **hydrogen embrittlement** of reinforcement bond loss as a result of the reduction of **rust** on the surface and the stimulation of alkali-aggregate reaction a result of the increased alkalinity in the region of the reinforcement. The relatively high current densities used in electrochemical chloride extraction make these effects potentially more severe than in the case of cathodic protection where the current densities are much lower, or electrochemical re-alkalization where the time is quite short.

5.13 Electrochemical re-alkalization

This process is similar to electrochemical chloride extraction, but the purpose is to generate alkaline hydroxyl (OH^+) cations from water and usually also to encourage alkaline metal ions (e.g. Na^+) to penetrate the cover concrete from an external source of suitable electrolyte (usually **sodium carbonate solution**) unless alkali reactive **aggregates** are present, in which case plain water is used. The advantage of introducing alkaline metal ions is that re-carbonation may be inhibited, though a barrier coating should give similar protection. The method is a relatively recent development (a version was first patented in 1992) but has become increasingly popular.

Temporary anode systems are used and the intrusion caused by the system is similar to **electrochemical chloride extraction**, but the process is complete in days rather than weeks.

Current densities range from 0.5 to about 2 A/m^2 of the concrete and **application voltages** from about 12 to 24 V. A **pH** of about 12 can be achieved after a few days. Testing whether sufficient alkalinity has been achieved is more difficult than with untreated concrete cover because the **phenolphthalein** colour change is complete at pH 10.4 and in this case there can be no assumption that the pH will be greater than 11 within a few millimetres of the test location (see Section 4.6.1).

Barrier **coatings** are usually used to protect the surface from re-carbonization. Where sodium carbonate solution has been used, the surface must be cleaned and the coating carefully chosen to avoid breakdown from remaining alkalis.

The possible undesirable **side-effects** are of the same kind as with other **electrochemical protection and repair methods**, but probably of a negligible magnitude because the total **charge** is quite small, say, likely to be less than 100 Ah/m^2 of concrete surface.

5.14 Over-cladding and sheltering

Over-cladding is the most expensive repair **option** but also the only one able to transform the exterior of a structure (usually a building) in both **appearance** and performance. Although over-cladding has been widely used for many years to **upgrade** buildings whether or not they are defective as a result of reinforcement corrosion, the question of whether the deterioration has been reduced to an acceptable level throughout the service life or merely covered up has been studied only recently [25].

Over-cladding increases the rate of **carbonation** and the average temperature of the concrete, both of which effects could be expected to increase corrosion, but can reduce the **relative humidity** of the concrete to a level as low as 65%, at which **corrosion currents** are very low. Adequate thermal insulation and ventilation in the cavity can ensure such conditions in heated buildings.

5.15 Other repair and protection methods

The treatment of reinforcement corrosion is very important commercially and it is not surprising that new **products and systems** for repair and protection are regularly brought onto the market. This guide concentrates on methods whose effectiveness has been demonstrated over several years of widespread use, but this should not be taken to mean that other methods are necessarily unsatisfactory provided they fulfil the basic principles of corrosion protection and repair. Clients contemplating the use of methods which do not have a history of widespread satisfactory performance should look for evidence of their efficacy in similar circumstances or undertake trials which will prove this.

6. Test methods and standards for repair materials and processes

Since the early 1970s, attitudes towards the repair and maintenance of buildings and structures have progressed from the practice of fixing defects by trial and error as they occur, to the application of engineering science and design to the task of maintaining buildings, structures and the infrastructure.

Nowhere has this development been faster than in the maintenance and repair of reinforced concrete. In terms of quantity at least, this subject dominated construction research in several developed countries for a number of years. In many of these countries distinct and specialized concrete repair industries developed as a result of owners' increased acceptance of the need to invest in the repair and maintenance of valuable fixed assets and infrastructure. Leading members of this industry contributed to the development of repair technology and practice as well as applying it to their own products and services, but the intense competition that results from the growth of any new market can also produce confusion and misunderstanding about the basic technical requirements that have to be fulfilled.

The codification and acceptance of common standards is as important to conscientious suppliers as it is to customers: it protects customers by providing them with an agreed specification for acceptable quality or performance, and protects suppliers by ensuring that their interests will not be damaged by undercutting or unreasonable customer demands.

A development which started in the mid-1990s has been the creation of a suite of European Standards covering the repair of concrete. The CEN 1504 Series: 'Products and systems for the protection and repair of concrete structures' is based on the experience and consensus of opinions of leading European experts and draws on a world-wide body of knowledge. Although the standards have been written specifically for repairs making use of proprietary products and systems, virtually all of the requirements and information in the series are equally applicable to any process of concrete protection and repair, regardless of the source of supply of the materials and systems.

The coverage of the series and the other standards it makes use of is still developing but ranges from the preparation of standard substrates for testing protection and repair materials, through apparatus and methods for testing the performance of products and systems under standard conditions, to their application and control on site.

Table 6.1 The BS EN 1504 series

Standard	Title	Scope	Status
BS EN 1504-1 to 10	'Products and systems for the protection and repair of concrete structures'	varies	Varies
BS EN 1504-1: 1998	Definitions	Definition of terms relating to 1504 standards.	Published
pr EN 1504-2	Surface protection systems	Increasing durability, maintenance and repair. Systems include hydrophobic impregnation, impregnation and coating.	Publicly available as a 'draft for public comment'
pr EN 1504-3	Structural and non-structural repair	Performance of repair grouts, mortars and concretes for structural and non-structural repair	Publicly available as a 'draft for public comment'
pr EN 1504-4	Structural bonding	Requirements, performance and conformity criteria for identification and safety of products for structural bonding of construction materials to a concrete structure.	Publicly available as a 'draft for public comment'
pr EN 1504-5	Concrete injection	Requirements, performance and conformity criteria for identification and safety of products for internal filling of cracks and cavities.	Publicly available as a 'draft for public comment'
Part 6:	Anchoring products	Requirements, performance and conformity criteria for identification and safety of products for anchoring reinforcing steel.	Publication date not known
Part 7:	Reinforcement corrosion protection	Requirements, performance and conformity criteria for identification and safety of products for preventing reinforcement corrosion, in particular by realkalization and electrochemical chloride extraction.	In preparation
pr EN 1504-8	Quality control and evaluation of conformity	Procedures for sampling, evaluation of conformity, marking and labelling of products and systems for protection and repair according to EN 1504.	Publicly available as a 'draft for public comment'
BS EN 1504-9: 1997	General principles for the use of products and systems	Basic considerations for specification of repair of plain and reinforced concrete.	Published
pr EN 1504-10	Site application of products and systems and quality control of the works	Requirements for substrate condition; storage, preparation and application of products and systems; quality control; maintenance; health and safety; environment.	Publicly available as a 'draft for public comment'

Table 6.2 Standards related to the BS EN 1504 series

Standard	Title or subject	Status
Eurocode 2 Parts 1–4: 1994	Design of concrete structures	Published
BS EN 206-1: 2000	Concrete: specification, performance, production and conformity	Published
BS EN 12696: 2000	Cathodic protection of steel in concrete: Part 1 Atmospherically exposed concrete	Published
	Realkalization for reinforced concrete	In preparation
	Chloride extraction for reinforced concrete	In preparation

Table 6.3 Topics covered by British and European Standards referred to in the BS EN 1504 series. The publication status of the standards varies

Adhesion after fatigue under dynamic loading during cure
Adhesion and elongation capacity of ductile injection products
Adhesion by slant shear strength
Adhesion by tensile bond strength
Admixtures for concrete, mortar and grout – Part 2: Concrete admixtures: Definitions, specifications and conformity criteria
Application overhead
Bond strength by pull-off
Building lime – Part 1: Definitions, specifications and conformity criteria
Capillary water absorption
Carbonation resistance
Cathodic protection of steel in concrete
Cement – Composition, specifications and conformity criteria – Part 1: Common cement
Chemical resistance
Compatibility on wet concrete
Compatibility with concrete
Compatibility with elastomer
Compatibility with steel
Creep in compression
Creep under tension
Determination of:
- adhesion concrete to concrete
- adhesion steel to steel
- air content of fresh concrete
- amine function
- ash by direct calcination
- carbon dioxide permeability
- carbonation depth in hardened concrete by the phenolphthalein method
- chloride content in concrete to be repaired
- coefficient of thermal expansion
- compatibility between overlay and anode
- compressive strength
- crack bridging ability
- creep in flexure
- durability of composite systems involving structural bonding agents
- epoxy equivalent
- fatigue under dynamic loading
- flow time by use of flow cups
- hydroxyl value
- injectability
- isocyanate content
- loss of mass after freeze–thaw salt stress – Testing of impregnated hydrophobic concrete
- modulus of elasticity in compression
- open time
- pot-life as function of temperature
- progressing in hardening (shore A or B)
- relative humidity
- resistance in sea water and/or high sulphate contents water
- shrinkage and expansion
- stiffening time
- tensile strength development for polymers
- thermal compatibility – Part 1: Freeze–thaw cycling with de-icing salt immersion
- thermal compatibility – Part 2: Thunder–shower cycling (thermal shock)
- thermal compatibility – Part 3: Thermal cycling without de-icing salt impact
- thermal compatibility – Part 5: Resistance to temperature shock
- viscosity
- volatile and non-volatile matter
- water-tightness
- wear resistance

Density of the adhesion bond to contact with water
Drying test for hydrophobic porelining impregnation
Electrochemical remediation of reinforced concrete; realkalisation
Environmental compatibility
Fire resistance
Flow Table test
Glass transition temperature
Granulometry size grading
Hardened concrete – Determination of depth of water penetration
Hardened concrete – Determination of pull out strength

Hardened concrete – Determination of ultrasonic pulse velocity
Infrared analysis
Linear shrinkage for polymers
Making and curing specimens for strength tests
Masonry cement – Part 1: Specification
Paint and varnishes – Cross cut test
Paints and varnishes – Coating materials and coating systems for interior masonry – Part 2: Determination and classification of water-vapour transmission rate (permeability)
Paints and varnishes – Coating materials and coating systems for exterior masonry – Part 3: Determination and classification of liquid-water transmission rate (permeability)
Paints and varnishes – Coating materials and coating systems for exterior masonry – Part 11: Methods of conditioning before testing
Paints and varnishes – Determination of resistance to abrasion – Part 2: Rotating abrasive rubber wheel method
Paints and varnishes – Evaluation of degradation of paint coatings – Designation of intensity, quantity and size of common types of defect – Part 1: General principles and rating scheme
Paints and varnishes – Evaluation of degradation of paint coatings – Designation of intensity, quantity and size of common types of defect – Part 2: Designation of degree of blistering
Paints and varnishes – Evaluation of degradation of paint coatings – Designation of intensity, quantity and size of common types of defect – Part 4: Designation of degree of cracking of coatings
Paints and varnishes – Evaluation of degradation of paint coatings – Designation of intensity, quantity and size of common types of defect – Part 5: Designation of degree of flaking
Paints and Varnishes determination of thickness
Paints and varnishes; determination of resistance to liquids; part 1: general methods
Paints and varnishes; falling-weight test
Pot-life of liquid systems: preparation and conditioning of samples and guidelines for testing
Preparation of steel substrates before application of paints and related products: visual assessment of surface cleanliness
Reactive functions of epoxy resins – thermogravimetry of polymers – temperature scanning method
Resistance to high chemical attack
Sampling fresh concrete
Setting time
Shrinkage – Grout for prestressing tendons – test method: Bleeding
Shrinkage – Grout for prestressing tendons – test method: Volume change
Shrinkage of polymer binders – Part 1: Determination of linear shrinkage for polymers and surface protecting systems
Soluble chloride content of fresh and hardened mortars
Special properties for aqueous gels
Specific weight – Pyknometer method
Specific weight – Immersed body method
Sprayed concrete for repair and upgrading of structures
Steel for the reinforcement of concrete
Suitability for application test: vertical and horizontal surfaces
Suitability for injection
Surface characteristics – Test methods – Measurement of skid resistance of a surface: Pendulum Test
Surface drying test – Ballotini method
Tensile strength, elongation and elastic modulus
Testing concrete – Cored specimens, taking, examining and testing in compression
Testing concrete – Determination of air content of fresh concrete
Testing concrete – Determination of compressive strength of test specimens
Testing concrete – Determination of consistency of fresh concrete – Vebe Test
Testing concrete – Determination of consistency of fresh concrete – Degree of compatibility
Testing concrete – Determination of consistency of fresh concrete – Slump test
Testing concrete – Determination of density of hardened concrete
Testing concrete – Determination of penetration of water under pressure
Testing concrete – Non-destructive testing – Determination of rebound number
Tests to determine the durability of structural bonding agents
Thermogravimetry of polymers – Temperature scanning method
Volumetric shrinkage
Water absorption and resistance to alkali test for hydrophobic porelining impregnation
Workability
Workable life

Bibliography

References

1. Sutherland J, Humm D and Chrimes M eds (2001). *Historic concrete: the background to appraisal*. Thomas Telford, London.
2. Wilkins NJM and Lawrence PF (1980). *Fundamental mechanisms of corrosion of steel reinforcement in concrete immersed in sea water*. Technical Report No. 6. Concrete in the Oceans, C&CA, Crowthorne.
3. *Marine survey of the Tongue Sands Tower* (1980). Technical Report No. 5. Concrete in the Oceans, C&CA, Crowthorne.
4. *The relevance of cracking in concrete to corrosion of reinforcement* (1995). Technical Report No. 44. Concrete Society, Crowthorne.
5. BS 2846 (1985–1997). *Guide to statistical interpretation of data*, Parts 1–7. BSI, London.
6. BS 6000 (1996). *Guide for the selection of an acceptance system, scheme or plan for inspection of discrete items in lots*. BSI, London.
7. *Guide to testing and monitoring the durability of concrete structures* (2002). Technical Guide No. 2. Concrete Society, Crowthorne.
8. Page CL and Treadaway KWJ (1982). Aspects of electrochemistry of steel in concrete. *Nature* **279**.
9. Vassie PR (1991). *The half-cell potential method of locating corroding reinforcement in concrete structures*. Application Guide 9. TRRL, Crowthorne.
10. Chess P and Gronwald F (1996). *Corrosion investigation: a guide to half-cell mapping*. Thomas Telford, London.
11. Currie RJ (1986). *Carbonation depths in structural-quality concrete: an assessment of evidence from investigations of structures and from other sources*. BRE, Watford.
12. BS prEN 104-865 (2000). *Determination of carbonation depth in hardened concrete by the phenolphthalein method*. BSI, London.
13. Pullar-Strecker P (1992). Reading the 'hallmarks' on concrete structures: a cautionary tale and message of hope. *Concrete*, Nov/Dec. Concrete Society, Crowthorne.
14. prEN 1504-10 *Site application of products and systems and quality control of works*, Part 10 of European Standard 1504 *Products and systems for the protection and repair of concrete structures*. BSI, London.
15. *Standard method of measurement for concrete repair* (1997). Concrete Repair Association, Aldershot (available for download at www.concreterepair.org.uk).
16. DD ENV 1504-9 (1997). *General principles for the use of products and systems*, Part 9 of European Standard 1504 *Products and systems for the protection and repair of concrete structures*. BSI, London.
17. Mays GC and Barnes RA (1995). The structural effectiveness of large volume patch repairs to concrete structures. *Proc Instn Civ Engrs, Structs Bldgs* **110**, 351–360.
18. Eurocode 2 (1994). *Design of concrete structures*, Parts 1–4. BSI, London.
19. BS EN 206-1 (2000). *Concrete: specification, performance, production and conformity*. BSI, London.
20. Standard BD 27/86 (1986). *Materials for the repair of concrete highway structures*. Highways Agency, London.

21. Advice Note BA 23/86 (1986). [Gives useful information on the design of repair mortar and concrete, including flowing concrete.] Highways Agency, London.
22. Sergi G, Seneviratne AMG, Maleki MT, Sadegzadeh M and Page CL (2000). Control of reinforcement corrosion by surface treatment of concrete. *Proc Instn Civ Engrs, Structs Bldgs* **140**, 85–100.
23. *Guide to surface treatments for the protection and enhancement of concrete.* Technical Report No. 50. Concrete Society, Crowthorne.
24. Broomfield J (1997). *Corrosion of steel in concrete – understanding, investigation and repair.* E&FN Spon, London.
25. Croker A, Mangat PS, Bougdah HN and Sharples S (2000). *Design guidelines for overcladding systems to maintain durability of the reinforced concrete building fabric.* School of Environment & Development, Sheffield Hallam University.

Further reading

Allen RT ed. (1998). *Concrete in coastal structures.*

Corrosion of steel in concrete, Part 1 *Durability of concrete structures,* Part 2 *Investigation and assessment,* Part 3 *Protection and remediation*. Digest 444 (2000). CRC Ltd, Watford.

Developments in durability design and performance-based specification of concrete CS 109. Concrete Society, Crowthorne.

Dhir RK ed. (2002). Challenges of concrete construction, Vol. 3 Repair, rejuvenation and enhancement of concrete. *Proceedings of an International Conference*, University of Dundee. Thomas Telford, London.

Forde MC ed. Structural faults and repair. *Proceedings of a Series of International Conferences on Structural Faults and Repair.* Engineering Technical Press, Edinburgh. Conferences are held regularly and the proceedings always include several useful papers on concrete repair.

Holland R (1997). *Appraisal and repair of reinforced concrete.* Thomas Telford, London.

Kay T and Walker M (2002). *Guide to the evaluation and repair of concrete structures in the Arabian Peninsular*. Special Publication CS 137. Concrete Society, Crowthorne. Much more than a specialist guide for arid regions; a useful guide to many aspects of concrete repair.

Kay T (1992). *Assessment and renovation of concrete structures*. Longman Scientific, Harlow. Thorough and detailed explanations of most aspects of concrete assessment, testing and repair.

Leeming MB and O'Brien TP (1987). *Protection of reinforced concrete by surface treatments.* Technical Note 130. CIRIA, London. Detailed information on the choice and applications of coatings for concrete.

Mallett GP (1994). *Repair of concrete bridges. State of the art review.* Thomas Telford, London. A compact sourcebook of assessment and repair methods, and case studies of repairs which have been carried out on concrete bridges.

Mays G ed. (1992). *Durability of concrete structures. Investigation, repair, protection*. E&FN Spon, London. Eleven chapters by authoritative specialists cover all aspects of the subject.

Neville AM (1981). *Properties of concrete.* Pitman. All there is to know about concrete as a material.

Page CL, Treadaway KJW and Bamforth PB eds (1990). Corrosion of reinforcement in concrete. *Proc Internat Sym on Corrosion of reinforcement in concrete construction, Soc Chem Ind.* Elsevier Applied Science, London. Fifty-two original papers from 16 countries cover basic research, useful applications and case studies. A mine of information for the serious student of the subject.

Rendell F, Jamberthie R and Grantham M (2002). *Deteriorated concrete: inspection and physicochemical analysis*. Thomas Telford, London. Detailed information on in-situ and laboratory testing and analytical techniques.

The route to a successful concrete repair (2001). Concrete Repair Association, Aldershot (available for download at www.concreterepair.org.uk). All the essential considerations for repair contracts compressed into three pages. Also a three-page appendix of repair materials.

Glossary and subject index

This glossary is intended to provide an alphabetical subject index, brief summaries of the main topics and definitions of terms used in this guide. Where appropriate definitions have been taken from relevant standards and other sources. Terms listed in the glossary have been emboldened in the text to support a topic search. The numbering refers to the section under which the subject will be mentioned and will be in bold if the subject is the section heading.

abseiling: 4.2

access equipment: **4.2**; 5.2: access equipment for initial investigations is usually limited to ladders, hoists and cradles; detailed investigations which need more access equipment are sometimes postponed until access equipment for repair is in place

accidental overloading: 5.1: actions which exceed the ultimate or serviceability limit state for which the structure was designed, e.g. impact, subsidence

accommodation: **4.2**: for initial investigations little or no accommodation is needed; for detailed investigations the client can often provide accommodation on site

active cracks: see cracks, cracking

active, in electrochemical sense: 2.1; 2.2.1; 4.5.1; 4.5.4: not passive; steel in a condition where corrosion is not prevented by the presence of a passivating layer: in some conditions where steel is active the corrosion rates may be minimal or corrosion may be absent for other reasons; achieving such conditions is often the objective of repair

additives: 2.2.2; 5.8.5: additions; fine inorganic material added to concrete to achieve certain properties; may be inert or may be reactive, e.g. pozzolanic material

admixture: 4.3.1; 5.8.5: a substance which is added in small quantities to a concrete mix to modify its properties

aggregate contamination: 3.2.3; 4.7: contamination, in this case with pyrites which causes rust staining and can cause corrosion as a result of erosion; or sea-dredged aggregates which can be contaminated with chlorides

aggregates, alkali reactive: 5.11; 5.12; 5.13: reaction between alkalis from cement and certain types of aggregate (principally silicates) can damage concrete

aggregates, composition and grading: 4.8.2; 5.8.5

aggregates, magnetic: 4.6.2: aggregates containing material which can interfere with cover meter readings

aggressive environments: 1: natural environments which can cause deterioration to be faster than normal; *or in this case* environments contaminated by industrial chemicals; *see* **introduction and scope**

alkaline environment, alkalinity (pH): 2.2.1; 4.6.1; 5.11; 5.12: in this case an environment more alkaline than **pH** 10–11, i.e. above the threshold for providing passivity; uncarbonated cement has a pH of more than 11; when fully carbonated the pH drops to about 8; **phenolphthalein** changes colour at about pH 9

angle-grinder, also masonry saw: 5.6.1: powered disk for cutting concrete; tends to leave a surface too smooth to bond without further treatment

anode external overlay, anode overlay: 5.11; 5.12: a conductor attached to the surface of concrete to act as an anode in an **electrochemical** process such as **cathodic protection**

anode, anodic: 2.1; 2.2; 2.2.2; 3.2.1; 3.2.4; 4.5.1; 4.5.5; 5.11: the **electrode** at which electrons leave a cell: it is positive in **electrochemical protection** and other processes which consume current and negative in cells which generate current *but it is always the site of any attack*

anode, incipient: 2.2.5: a potentially anodic area which is sacrificially protected by a strongly corroding neighbouring anode

anodic control: 2.2; **2.2.4**; 5.4: control of corrosion by preventing potentially anodic areas from functioning, e.g. by applying a barrier coating to them

appearance: 5.4; 5.6.2; 5.10; 5.14: protection and repair can often provide an opportunity to improve appearance with surface coatings

application voltages: 5.11; 5.12; 5.13: the voltage at which electrochemical methods of protection and repair are applied, usually 24 V or less to avoid injury to people or animals

arbitration: 3.1: the process of resolving a formal dispute by an arbitrator who has knowledge of the subject

assessment: 3.1; 4.2; 4.7; 4.11: evaluation of condition, properties or **deterioration**, usually leading to repair **decisions**

assumed bar diameter: *see* reinforcement, assumed bar diameter

bar chart: 4.6.3: **histogram**; a chart of bars whose areas represent the relative frequencies of values; useful for a visual check of whether values represent more than one statistical population

black rust: 2.2; 3.2.4; 4.7: black (or initially green) **corrosion products** of iron formed where little oxygen is available; sometimes magnetite

blasting with fine jet high-pressure water: 5.6.1: a useful method for cutting concrete

blasting with grit: 5.6; 5.6.1; 5.7.1; 5.8.4; 5.10: air blasting with an abrasive, used to clean surfaces; grit blasting can cut reinforcement

blasting with high-pressure water: 4.9; 5.6; 5.6.1; 5.7.1; 5.8.4: blasting to remove concrete with water at a pressure above 50 MPa

blasting with a low-pressure water jet: 5.7.1: blasting with water at a pressure up to 15 MPa for cleaning

bleeding: 3.2.1; 5.8.5: watery liquid rising to the surface of mortar or concrete

bond: 5.6; 5.6.1; 5.7.1; 5.7.2; 5.8.1; 5.8.4; 5.12: permanent adhesion between **substrate** or parent concrete and repair material or reinforcement

bonding coat: 5.8.4: coating sometimes used as an interface between layers with the intention of promoting **bond**

brief: **3.1**; 3.3: in this case, written instructions for the investigation work agreed with the **client**: for **initial investigations** it may be a simple letter; for **detailed investigations** it should be a comprehensive document. A written brief should be insisted on by both parties; a checklist of suggested headings is included in the text

bulk sample: 4.7: initial sample from which representative portions are taken for **chemical analysis**, retention, or confirmation; bulk samples should weigh at least 25 g

calcium chloride: 4.3.1; 4.7: an admixture widely used in the 1950s and 1960s to accelerate the rate of gain of strength of concrete, especially in cold weather; it causes corrosion and was progressively banned from 1965 to 1985, first for prestressed and then for reinforced concrete

calcium oxide, hydroxide: 2.2.1: the main alkaline constituent of cement; its low solubility (*ca.* 2%) results in alkalinity being preserved for some time during the carbonation process, especially in rich mixes

carbon dioxide penetration, carbonation: 2.2.1; 2.2.6; 3.1; 3.2.1; 3.2.3; 3.2.4; 4.5.4; **4.6.1**; 5.4; 5.5.4; 5.5.5; 5.6; 5.6.1; 5.6.2; 5.7.1; 5.10; 5.14: the reaction of atmospheric carbon dioxide with water and (mainly) calcium hydroxide in concrete to reduce the **alkalinity** of the pore fluid from more than **pH** 11 to about pH 8

carbonation coefficient: 4.6.1; 4.6.3; 5.6.2: diffusion coefficient; the **K-value** which is related to the speed at which carbon dioxide penetrates concrete in the equation $d = K\sqrt{t}$, where d is the depth in millimetres and t the time in years; values of K range from about 0.5 to 3 for high-quality concrete, e.g. precast concrete, to about 9 for low-quality in-situ concrete with an average of about 5

carbonation front: 4.6.1: the boundary between **carbonated** and uncarbonated concrete; in practice usually the front identified by **phenolphthalein**, which can be up to about 10 mm shallower than the front at which reinforcement is protected; *see* **alkalinity**

cathode, cathodic: 2.1; 2.2; 3.2.1; 3.2.4; **5.11**; 5.12: the electrode at which electrons enter an electrolytic cell: its polarity depends on whether the cell is generating current (positive) or consuming current (negative) but *it is never the site of attack*; *see* **anode**

cathodic control: 2.2; **2.2.3**; 3.2.4; 5.4: preventing corrosion by depriving all potentially cathodic areas of the oxygen needed to keep them active, i.e. keeping them in a state of **polarization**; occurs in saturated self-contained units which are not electrically connected to other units which have oxygen access

cathodic prevention: 5.11: installation of a **cathodic protection** system in a new structure during construction; the system is either left unpowered until needed or powered at a very low level; used where long-term durability is critical and environmental conditions are potentially aggressive

cathodic protection: *see* **electrochemical methods of repair**

cell: 2.1; 5.11: electrolytic cell; the unit formed by an **anode** and a **cathode** in an **electrolyte**

cement content and type: 4.7; 4.8.2; 5.8.5; 5.9: the proportion of cement in concrete; usually refers only to the Portland cement; sulphate-resisting cement and ASTM Type V contain low proportions of tricalcium aluminate and have a reduced ability to bind chloride contamination

cementitious materials: 5.6; 5.8.1; 5.8.3; 5.8.4; 5.9: material containing hydraulic (Portland) cement as the main active constituent

cementitious products, polymer modified: 5.6: 5.8.3; 5.8.4; 5.9: cement-based mortar or concrete modified by the addition of a polymer

charge, total: 5.12; 5.13: the total electrical current applied during **electrochemical protection and repair methods**

chemical analysis: 3.2.1; 3.2.4; 4.5.1; 4.7; 5.6.3: here, quantitative analysis to determine the proportion of a chemical agent (e.g. chloride ion in concrete). **Volard** or **potentiometric titration**; 4.7: or indirect comparative methods like X-ray fluorescence are normally used; **X-ray fluorescence**: 4.7: a method of quantitative analysis which compares the X-ray fluorescence of samples under test with that of samples with known concentrations of the contaminant; used for rapidly evaluating levels of **chloride contamination**; *also* **Hach, Quantab**: 4.7: field tests designed to determine the chloride content of aggregates; they can be adapted as field tests for use on samples of hardened concrete

chloride, chloride contamination: 2.2; 2.2.1; 2.2.5; 2.2.6; 3.1; 3.2.1; 3.2.3; 3.2.4; 4.5.1; 4.5.6; **4.7**; 4.8.2; 4.9; 5.1; 5.4; 5.5.1; 5.5.4; 5.6; 5.6.2; **5.6.3**; 5.7.1; 5.10; 5.11; 5.12: the chloride ion, usually originating from **salt** or **calcium chloride**, which if present at more than 0.4–0.5% Cl ion as by weight of cement for mixed-in chloride or more than about 0.2% for penetrated chloride, can cause corrosion by catalytic action; **cement content and type**, and **carbonation**, affect the corrosion thresholds; *also* **chloride penetration**: 4.5.1; 4.7; 4.8.2; 5.6.3; 5.10: the diffusion of chloride ion into concrete from an external source, creating a chloride gradient which decreases with distance from the point of entry

chlorides in the mix: 4.7; 5.6.3: *cf.* **chloride penetration**

choice of protection and repair methods: 5.4: a checklist of repair options with their advantages and disadvantages is included in the text

Cintride-tipped bit: 4.5.3: a special drill bit designed for drilling difficult metals, e.g. stainless steel, cast iron and high-yield steel

client: 3.1; 3.2; 3.3; 3.4; 4.1; 4.2; 4.3; 4.11; 5.1; 5.4: used here to denote the person or organization with authority to commission inspection and assessment and to make **decisions** on repair options

coatings and surface treatments: **2.2.4**; 3.2; 5.4; 5.6.2; **5.10**; 5.11; 5.13: layers applied as liquid which hardens or powder which is fused; barrier coatings applied to concrete; *also* **penetrating coatings**: 5.10: impregnating liquid products which penetrate the concrete and block the pore system; and **surface treatments**: 5.10: penetrating hydrophobic liquids which affect the surface tension of the pores (*see* **silanes and siloxanes**); can be **solvent-based**: liquid coating which is carried in a solvent which subsequently reacts or evaporates; or **water-based**: liquid coating which is carried in water which subsequently reacts or evaporates; *also* **barrier coatings**: 2.2.4; 5.10; 5.13: to resist the penetration of carbon dioxide or water, or decorative coatings to improve appearance, especially after **patch repairs**; if they are breathable, such coatings can allow concrete to become drier; *also*, **carbon dioxide resistant coatings**: 5.6.2; 5.10: barrier coating which reduces the rate of transmission of carbon dioxide because of the relatively large size of carbon dioxide molecules, such coatings can be porous to water vapour and allow concrete to become drier; *also*, **crack bridging elastomeric coatings**: 5.10: barrier coating able to span cracks of limited width without rupturing; *also*, **electrically insulating coatings**: 5.4: coatings which can be applied to reinforcement to isolate potentially **anodic** areas for **anodic control**

concrete permeability constant: 4.6.1; 4.6.3: in this case, the '**carbonation coefficient**' or '**K-value**'

concrete quality: 3.2.1; 3.2.4; 4.3.1; 4.6.1; 4.8.1; 4.8.2: durability requirements based on water/cement ratio and minimum cement content; *also* **concrete density:** 4.8.2: here, the absence of voids; also **concrete permeability**: 4.8.2: *also* **concrete, minimum cement content**: 4.8.2; 5.8.5: *also* **concrete potential strength**: 4.8.1; 4.8.2: compressive strength measured by testing cubes made for the purpose

concrete repair mixes, proprietary dry-bagged: 5.8.2; 5.8.3; 5.8.4; 5.8.5; 5.9: repair materials provided in a ready-to-use form by a supplier; *see* **proprietary products and systems**

concrete, hardened: 3.2.1: concrete by the time it has developed some strength

concrete, in-situ: 4.6.1; 4.7; **4.8**: site-poured concrete, i.e. not precast; *also* **tests on in-situ concrete**: **4.8**

conductive paints: 5.11: coating containing electrically conductive pigment or filler which forms a conductive layer when cured or hardened; used in **anode systems** for **cathodic protection**

conductivity: 2.2.2: *see* **resistivity**

connecting screw: 4.5.3: here, screw used to connect the positive lead in surface **electrode-potential** measurement

contract: 4; 4.1; 4.3.1; 5.2: contracts and tenders

copper/copper sulphate half-cell: 4.5.2; 4.5.4; 4.5.5: a probe consisting of a copper electrode in a saturated solution of copper sulphate, widely used on site in **surface electrode potential** measurement

core samples: 4.3; 4.3.1; 4.7; 4.8.1; **4.8.2**: cylindrical specimen cut from the hardened concrete in a structure; *also* **core strength**: 4.8.1; 4.8.2: strength of concrete in axial compression or indirect tension measured on a core cut from a structure; *also* **core, location and orientation**: 4.8.2: permanent markings made within and adjacent to the circumference of the area to be removed to identify the core and the core hole

corrosion current: 2.2; 2.2.2; 4.5.2; 4.10; 5.4; 5.11; 5.14: the electric current flowing between the **anodic** and **cathodic** areas when reinforcement corrodes; *also* **corrosion current measurement**: 4.10: linear polarization method

corrosion of steel in concrete: 2.2

corrosion, general: 2.2; 3.2.1; 3.2.4; 4.6.2: uniform rusting which is free from visible **pitting**; usually occurs when reinforcement has lost **passivity** in relatively dry conditions; separate **anodes** and **cathodes** are not visible

corrosion, pitting: 2.2; 3.2.1; 3.2.3; **3.2.4**; 5.7.1; 5.7.2: the visible pitting at concentrated **anodic** areas where corrosion is driven by larger **cathodic** areas some distance away in wet **chloride contaminated** concrete; *also* **corrosion products**: 2.2; 3.2.3; 3.2.4; 4.7: **soluble iron compounds**, partially oxidized compounds (**black rust**) or fully oxidized red rust

cover meter: 3.2; 3.2.1; 4.2; 4.5.3; 4.6.2; 4.6.3: an electromagnetic device used to measure the distance of reinforcement below the concrete surface

crack inducers: 3.2.1; 5.5.3: inclusions or interruptions which encourage concrete to crack; can include transverse reinforcing bars

crack injection: 5.5.3; 5.5.5

crack width classification: 3.2.1; 4.3: classifying cracks from 0.05 to more than 1.0 mm according to their width and visibility

cracks, cracking: 2.2; **3.2.1**; 4.3; **4.4**; 4.5.4; **5.5.1**; **5.5.2**; 5.5.3; **5.5.4**; 5.9; 5.10: **active cracks**: 4.4; 5.5.3; **5.5.5**; 5.10; 5.12: cracks which open and closes and behave as unintended movement joints; during repair these cracks must be sealed as movement joints; *also*, **controlled cracks in tension zones**: 3.2.1; 5.5.3: cracks whose width is controlled by reinforcement: codes usually specify a limit of 0.3 mm for most exposure classes but 0.2 mm or less for water-retaining structures or exposure to aggressive chemical environments [eur2; 4.4.2.1]; *also* **plastic shrinkage cracks**: 3.2.1; **5.5.1**: cracks formed when recently poured concrete shrinks as a result of loss of water; cracks are usually closed at both ends and roughly perpendicular to the longest dimension; *also* **plastic settlement cracks**: 3.2.1; **5.5.2**: cracks which can form locally near the tops of **pours** where the settlement of (usually very workable) concrete is held up by reinforcement; they tend to fill with **bleed** water and can leave the reinforcement unprotected from corrosion; *also* **longitudinal cracks**: 3.2.1; **5.5.4**: in this case cracks which form in the same direction and location as the underlying reinforcement; often a sign that the reinforcement is corroding; *also*, **thermal contraction cracks**: 3.2.1; 5.9: cracks which usually form soon after the concrete has hardened and passed the peak of the exothermic reaction; they result from restrained thermal contraction

cracks, plastic: *see* **cracks, cracking**; cracks formed before the concrete has fully hardened; they include thermal, shrinkage and settlement cracks

cracks, transverse: 3.2.1; **5.5.3**: here, cracks transverse to the reinforcement being considered; they are unlikely to be caused by corrosion or to cause it though lower layers at right angles may be affected

cumulative frequency diagram: 4.6.3: an ogive; a 'less than' or 'more than' curve representing the proportion of a statistical population that has more or less than a given value; grading curves for sand are a familiar example

curing: 5.9: maintaining sufficient moisture to allow continued hydration of the cement, especially at the surface (which can include the substrate in the case of repairs); best achieved with wet curing; *also* **curing membrane**: 5.9: an impermanent coating applied to newly-compacted concrete to prevent loss of water by evaporation

current density: 5.11; 5.12; 5.13: current in relation to area of reinforcement or of concrete for **electrochemical repair and protection methods**

damage: 3.1: an unacceptable condition caused by deterioration or exterior action

decisions, initial and strategic: 3.1; **3.4**

decisions, repair: 4.8.2; 4.11; 5.1; **5.4**

defect, defective: 3.1; 3.2.1; 4.3; 5; 5.1: an unacceptable condition which may be built in or may be the result of deterioration or damage

delamination and spalling: delamination: 2.2; 3.2.1; **3.2.2**; 3.2.4; 4.3; **4.4**; 4.5.4; 4.6.2; 4.8.2; **5.5**: separation of a surface layer from the body of concrete without being completely detached; **spalling**: 3.2.1; **3.2.2**; 3.2.3; 3.2.4; 4.3; **4.4**; 4.6.2; **5.5**; 5.10: part of a surface layer which has become completely detached from the body of the concrete: follows **delamination**

depth of carbonation: 4.4; 4.5.3; **4.6**; 4.6.1; 4.6.3; 4.8.2; 5.6.2 the depth to which the **carbonation** reaction has reduced the **alkalinity** of the concrete to less than the corrosion threshold of pH 10–11. It is d in the equation $d = K\sqrt{t}$; *see* **carbonation front**

depth of cover measurement and interpretation of carbonation and cover depth results: 4.3.1; 4.5.3; **4.6**; 4.6.1; **4.6.2**; **4.6.3**; 5.6.1: measurement of cover depth either directly or with a cover meter

depth of removal: carbonated concrete: 5.6.2

depth of removal: chloride-contaminated concrete: 5.6.3

'desalination': *see* **electrochemical protection and repair methods**

design assumptions: 4.2: the performance requirements intended to be achieved by the designer

design life: the intended useful period of service under expected conditions of use; *see* **service life**

design of recasting mixes: 5.8.5

deterioration: 3.2; 4.1; 4.3; 4.3.1; 4.11; 5.1; 5.6.1; a progressive loss of condition or performance

disputes: 3.3; 4.1; 4.2; 4.6.2; 4.7; 4.8.2; 4.11: here, disagreements between parties which can lead to claims; normally resolved by, mediation **arbitration** or **litigation**

drawings, accommodation, equipment and access: 4.2

drilling reinforcement (for electrical connections): 4.5.3: making an electrical connection with the reinforcement is an essential part of **surface electrode potential** measurement

durability: 4.8.1; 5.1: the combination of qualities needed to meet long-term performance requirements; *see* **design assumptions**

dust, hydrating: 5.6: dust formed when concrete is removed by dry methods can contain enough unhydrated cement to set when moisture is present

economic and strategic issues: 3.4; 4.11: non-technical considerations often decide the action the client will take when the condition of a structure has been assessed

elastic modulus: 4.10; 5.8.1; 5.8.5: ultrasonic pulse velocity can be used to estimate this property non-destructively; *see* **other tests**

electrochemical chloride extraction: *see* **electrochemical methods of repair**

electrochemical protection and repair methods: 2.1; 2.2; 2.2.5; 2.2.6; 4.4; 5.4; 5.5.1; 5.5.2; 5.6; 5.6.2; 5.6.3; **5.11**; **5.12**; **5.13**: methods which reduce or avoid need for mechanical repair by making use of reactions which take place between the charged **ions** in solutions, usually under the influence of applied power; **cathodic protection**: 2.1; 2.2; **2.2.5**; 5.4; 5.6.3; **5.11**; 5.12: keeping reinforcement cathodic (in relation to an external anode overlay) throughout the service life; can be by impressed current or by **sacrificial anode**; *also* **electrochemical re-alkalization**: 2.1; 2.2.6; 5.12; **5.13**; introduction of alkaline material or generation of hydroxyl ions to increase **alkalinity**; a one-off treatment which requires a short term application of power through a temporary external anode tank; may require **coating** to prevent loss of alkali by leaching; *also* **electrochemical chloride extraction**: 2.1; 2.2.6; 5.4; 5.6.3; **5.12** (sometimes called 'desalination' or electrochemical chloride removal); electrochemical extraction to reduce chloride ion concentration near reinforcement; a one-off treatment which requires a short-term application of power through a temporary external anode tank; may require **coating** to prevent re-penetration of chloride

electrochemical re-alkalization: *see* **electrochemical protection and repair methods**

electrochemical series: 2.1; 5.11: the relative order of reactivity of metals forming electrodes and therefore the ranking of their potentials under standard conditions

electrode potential mapping: 3.2.1; 3.2.4; 4.4; **4.5.6**; 4.7: marking contours of equal electrode potential on the concrete surface or on a drawing of it; used to pinpoint anodic areas and therefore (especially) to locate areas where chloride contamination may be concentrated

electrode potential, surface: 2.1; 3.2.4; **4.5**; 4.5.2; 4.5.5: the electrical potential of the reinforcement in relation to the concrete surrounding it; values are quoted with reference to a specific type of **half-cell** and are normally negative

electrode potential: 2.1; 4.5.1; 4.5.6: the potential of a **half-cell**; the **electrochemical series** gives the values of half-cell potentials under standard conditions in relation to hydrogen

electrode: 2.1; 4.5.2: here, a conductor in a conducting solution (**electrolyte**) or an anode used in **electrochemical protection and repair methods**

electrolyte: 2.1; 4.5.2; 5.12: an electrically conducting solution containing **ions**, e.g. those formed when salts dissociate on going into solution

electrolytic action or process: 2.1; **5.11**; **5.12**; 5.13: a process achieved by **electrolysis**, here by the passage of an electric current through an **electrolyte**

evidence: 3.1; 3.3; 4.1; 4.2; 4.6.2; 4.8.2; 4.10; 4.11: fact which can be proved, or an expert report written by an appropriately qualified person; used to support a claim in a dispute (expert reports are 'privileged', i.e. confidential to clients unless exchanged, and can therefore be rejected by them if they do not suit their purpose, but the opposition can subpoena the expert if they discover who it is)

faulty construction: 5.1: built-in defects

ferric oxide, gamma Fe_2O_3: 2.2.1: in this case the oxide coating which forms on reinforcement in highly **alkaline** conditions and renders it **passive**

fire damage: 4.9; 5.1

flowing concrete: 5.8.5: repair concrete cast in formwork and compacted by gravity

formwork: 4.3.1; 5.8.4; 5.8.5: special formwork for repair by recasting; *also* poor fixing of formwork can cause deterioration as a result of leakage at joints or loss of cover depth

gloved hand: 5.8.3: a method of placing and compacting hand-applied repair materials; the success of the repair is critically dependent on conscientious workmanship

guarantees: 5.8.2

half-cell: 3.2; 4.2; **4.5.2**; **4.5.4**; 5.11: in this case, an electrode in a solution of known concentration formed into a probe which is used to contact a concrete surface in surface electrode potential measurement; *also* **taking and interpreting half-cell readings**: **4.5.5**: *also* **checking the half-cell**: 4.5.4

honeycombing: 4.3; 4.3.1: the open texture resulting from a loss of grout, cement paste or fines from fresh concrete, often at formwork joints

hydrogen evolution and embrittlement: 5.11; 5.12: hydrogen evolution caused by 'over-voltage', i.e. by the additional potential at the cathode above that theoretically required to run the process, can cause embrittlement of some types of steel, especially high-yield steel, as a result of diffusion of hydrogen in the steel; this phenomenon limits the use of electrochemical processes in prestressed concrete

hydrophobic surface treatment: *see* **coatings and surface treatments**

impressed current: 5.11: the system of cathodic protection using current from an external power source, usually a transformer-rectifier system

improvement and upgrading: 3.4; 5.14: *see* **options** and **strategy**

in-situ concrete: *see* concrete

insurers: 3.3: those who write policies for the insurance insist that they handle claims for any recovery of costs from third parties: it is insurers, not the client, who are likely to initiate action against investigators

introduction and scope: 1

investigation and testing, purpose of detailed: 4.1

investigation, detailed: 3.1; 3.4; **4**; **4.1**; 4.2; 4.3.1; 4.4; 4.11; 5.1; 5.2; 5.6.2: investigation and assessment intended to establish the causes and quantity of deterioration in enough detail to enable the client to make well-founded decisions on the action to take

investigation, initial, and assessment of concrete structures: **3**: *also* **initial and strategic decisions: 3.4**: *also* **scope and brief for the initial investigation**: **3.1**; 3.2.1; 3.3; 4.2; 4.3.1

investigations, vulnerable locations checklist: where to look for corrosion damage and how to recognize it: 3.2

ions: 2.2.2; 4.5.2; 5.11; 5.12: in this case, the charged atoms or groups formed by the dissociation of molecules when they go into solution; they are the carriers of the **corrosion current** in concrete; *see* **charge**

joints: 4.4; **5.5.5**; 5.6.3; 5.12: includes construction joints which are not intended to leave a discontinuity, movement joints designed to allow for shrinkage, thermal expansion or settlement, and cracks which act as movement joints; can be locations for chloride intrusion

K-value: 4.6.1; 4.6.3: the '**carbonation coefficient**' in the equation $d = K\sqrt{t}$, where d is the depth in millimetres and t the time in years

lime: 4.6.1: calcium oxide or hydroxide

limestone: 4.6.1: rock composed mainly of calcium carbonate; can give temporarily misleading results in the **phenolphthalein** test

litigation: 3.1: the process of trying a **dispute** in a court

load-bearing: 5.6; 5.8.1; 5.8.5

log book: 4.3: an 'owner's manual' for a building or structure

maintenance after repair: 5.1 5.6.2; 5.10: recurrent or continuous measures which provide protection; *see* **monitoring and inspection**

making the connection with the reinforcement: 4.5.3; *see* **drilling reinforcement**

masonry saw: 5.6.1: *see* **angle grinder**

microcracking: 5.6: hairline cracks formed when concrete is removed with **power hammers**; can interfere with bond of replacement materials; *see* **water blasting**

monitoring, inspection and re-inspection: 3.3; 3.4; 4.5.2; 5.1; 5.4; 5.10: regular or periodic inspection or testing to identify changes in condition; *see* **maintenance after repair**

mortar, proprietary dry-bagged: *see* **concrete repair mixes**, or **proprietary materials, products and systems**

mortar, trowel or hand-applied: 5.8.3: here, cementitious or polymer-modified mortar designed to be placed and compacted with a trowel or **gloved hand**

negligence: 3.3

occupied sites: 4.2; 4.4; 4.7; 5.3; 5.4; 5.11: here, buildings and structures which are in use; *see* **occupiers**

occupiers: 4.2; 5.3; 5.4: repair often requires work to be done on occupied premises where some operations may have to be restricted or avoided altogether to avoid causing disturbance or nuisance; this can be important in considering **options**

options and strategy: 3.1; 3.4; 4.1; 4.4; 4.7; 4.11; 5; **5.1**; 5.4; 5.6; 5.10; 5.14: considerations for the **client** when making **decisions** following an assessment of **deterioration**; the client's choices for action (or lack of it)

other methods: 5.15

other tests and measurements: 4.10

out of date surveys: 3.3

overcladding and sheltering: 5.4; 5.6.2; **5.14**: adding an external shell or cladding to shelter the structure and sometimes also to improve its appearance and performance

oxygen: **2.2.3**; 3.2.1; 3.2.4; 4.5.5; 5.11: oxygen is necessary at the cathode to sustain corrosion; *see* **cathodic control**

parent concrete: *see* **substrate**

passive, passivity: 2.2; **2.2.1**; 2.2.2; 4.5.1; 4.5.4, 4.5.6; 5.4; 5.7.1; 5.8.1: not **active**; in a condition where steel is protected from corrosion by a passive oxide film (**gamma ferric oxide**) which forms in uncontaminated **alkaline** conditions (pH > 10–11)

patch edges, treatment of: 5.6.1: cutting patch areas to the correct angle to reduce bond stress; *see also* **removing concrete cover**

patch repair: 5.4; 5.6.2: small areas of replacement concrete or mortar

personal injury or damage to property: 3.2; 3.2.2; 4.4; 5.6: *see* **safety**

pH: 2.2.1; 4.6.1; 5.13: a measure of the degree of alkalinity on a scale of 7 (neutral) to 14 (alkaline); chemically the log of the reciprocal of the concentration of the hydrogen ion

phenolphthalein indicator solution: 3.2; 3.2.1; 4.2; 4.4; 4.6.1; 5.13: an indicator solution which changes from colourless to pink at a pH of about 9 (the exact range of colour change is pH 8.3–10.4); more visible on concrete than alternative indicators

photographs: 3.2.1; 4.2; 4.3; 4.4; 4.5.6; 4.8.2

pitting: *see* **corrosion, pitting: 3.2.4**

plasticiser: *see* admixture

polarization: 2.2; 2.2.3; 3.2.4: depriving potentially cathodic areas of enough oxygen to keep them inactive; *see* **cathodic control**

polymer modifiers: 3.8.3; 5.8.4; 5.8.6; 5.9; *see* **cementitious products, polymer modified**

pore fluid in inter-connected capillaries: 2.1; 2.2.2; 5.11: pore solution in inter-connected capillaries which accounts for the conductivity of concrete; *see* **resistivity**; *see also* **calcium oxide, hydroxide**: 2.2.1

pour: 4.3.1; 5.8.4; 5.8.5: the in-situ concrete which is placed in one operation; the whole unit can sometimes be treated as a single entity for statistical purposes; *see* **sampling**

power hammers: 5.6: the usual way of removing concrete, especially in small areas; can result in **microcracking** which interferes with **bond**

precast concrete: 4.7: tends to have a low **K-value**; can be a source of contamination from **calcium chloride**

preventive action: 3.3

production of electricity by electrochemistry: 2.1

properties of replacement materials: 5.8.1

proprietary materials, products and systems: 5; **5.8.2**; 5.8.3; 5.8.4; 5.8.5; 5.9; 5.10; 5.15: repair materials and systems provided in a ready-to-use form by a supplier; also **concrete repair mixes**

public and client relations: **5.3**; *see* **occupied sites and occupiers**

pull-off tests: 5.6; 6

radar scanning: 4.10

rebound hammer: **4.8.1**: a spring-loaded bolt which impacts on the hardened concrete surface and rebounds to an extent which increases with the strength (or elastic modulus) of the concrete; a Schmidt hammer

recasting with concrete: **5.8.4**; **5.8.5**: repair concrete or mortar cast in place using formwork; may be compacted by vibration, or in the case of **flowing concrete** by gravity

recording defects and test results: 4.2; **4.3**

reference electrode: 4.5.2: a standard electrode used to calibrate practical **half-cells**

reinforcement bond loss: 5.11; 5.12; also **rust, reduction**

salt, salt contamination, salty environments, sea-water: 2.2; 2.2.3; 3.2.1; 3.2.4; 4.7; 5.5.1; 5.5.3; 5.5.4; 5.5.5; 5.10: in this case common salt (NaCl) is implied

samples of drill-dust: 4.7; 5.6.3: samples of concrete for analysis obtained by drilling groups of holes in the structure with a large bit and collecting the dust; *see* **bulk samples**

samples, sampling scheme: **4.3.1;** 4.6.3; 4.7; 5.5.1; 5.6.3: a scheme designed to represent a particular condition, e.g. a worst-case or best-case condition, or random samples which can be treated statistically; *see* **statistically valid values**

saturated concrete: 2.2.3; 3.2.4; 4.5.5; 4.7: in this case concrete where there is insufficient oxygen to allow the cathode of a corrosion cell to function; *see* **cathodic control** and **polarization**

Schmidt hammer: 4.8.1: a design of **rebound hammer**

sea-water: 2.2.3; 3.2.1; 3.2.4; 4.7: sea-water contains about 1.6% chloride in temperate regions

service life: 2.2; 3.1; 3.2.4; 4.6.1; 5.1; 5.4; 5.6.2; 5.11: the actual period for which the intended performance is achieved

shrinkage-compensating admixture: 5.8.5: an admixture which causes concrete to expand slightly while hardening thus off-setting shrinkage; it is important to avoid over-compensation

shut-down: 4.4; 5.3: *see* **occupied sites**

side-effects, including long-term effects: 5.11; 5.12; 5.13

silanes and siloxanes: 4.5.4; 5.10: silicon compounds applied to hardened concrete to make it resistant to the absorption of water or salt solution: used to protect bridges from salt run-off; some types form a chemical bond with concrete and are used as primers to improve the adhesion of subsequent coatings; *see* **coatings** and **surface treatments**

silver/silver chloride cells: 4.5.2: **half-cells** made from silver in solid silver chloride; often used where cells are permanently embedded because of their long-term stability; silver/silver chloride cells based on saturated potassium chloride have potentials 98 mV less negative than copper/copper sulphate cells

site batched material: 5; 5.8.2: site mixed material; here, concrete or mortar mixed on or near the construction site by the user

smoothing coat, coating: 5.10: also called fairing coat; a **coating** applied to a concrete surface to fill voids and cracks before final coatings are applied; *see* **coatings**

sodium borate solution and sodium carbonate solution: 5.12; 5.13: reagents used in **electrochemical protection and repair methods**

soluble iron compounds: 3.2.3; 3.2.4; 4.7: in this case usually ferrous chloride before it has become oxidized to insoluble iron compounds; *see* **corrosion products**

spacers: 4.3.1: plastic or concrete disks etc clipped or tied onto reinforcement to keep it at the distance from the formwork required to provide the minimum specified cover; low cover depth is the most common cause of reinforcement corrosion

spalling: *see* **delamination**

sparks from burning steel: 5.6.1: a useful visual indication that cutting, for example with an **angle grinder**, has reached the reinforcement

specification and contracts for repair: 3.1; 4; **5.2**; 5.3

sprayed concrete or mortar: **5.8.6**; 5.11: concrete or mortar applied under pressure through a nozzle; *also* **dry-sprayed process**: spraying where the dry materials and water are combined at the spray nozzle; *also* **wet-sprayed process**: spraying where the mixed materials including water are combined before they reach the spray nozzle

statistically valid values: 4.3.1; 4.6.3: values which can be analysed by statistical methods to allow inferences to be made; for example values from sufficient **samples** taken at randomized locations

strain gauge, demountable mechanical: 4.4: a strain gauge which can be held across studs fixed either side of a crack to measure its movement

structural assessment: 4.2; 4.9; 5.6: evaluation of the load-bearing capacity of the structure

structural load capacity: 4.9: the load which can be applied without exceeding the design limit states

structural loading: 3.2.1; 5.6; 5.8.1: here, loading which has caused cracking whether or not it exceeds structural load capacity

structures which are in use: *see* **occupied sites**

substrate: 5.8.1; 5.8.3; 5.8.4; 5.8.5: **parent concrete**

superplasticizer: 5.8.5: admixture which greatly increases the workability of concrete for a short time

supervision: 5.2

support, temporary: 5.6: here, falsework which resists all actions which may occur before or during the repair process

surface blemishes: 5.10

surface tensile strength tests: 5.6

surface treatment: 3.2; 5.10: *see* **coatings and surface treatments**; *also* **silanes and siloxanes**

surface water absorption: 4.10: ISAT; an empirical test to determine how quickly water is absorbed by a concrete surface

Swedish SA $2\frac{1}{2}$: 5.7.1: standard of cleanness (steel) according to ISO 8501-1

thermal contraction and expansion: 3.2.1; 5.5.5; 5.8.1; 5.8.5; 5.9

ultrasonic pulse velocity: 4.10: the speed of ultrasonic pulses through concrete; it is related to elastic modulus and therefore to strength

understanding corrosion: 2

unhydrated cement particles: 4.6.1; 5.6: particles of cement in mature concrete which are not fully hydrated; exposure of the grains by cutting concrete can cause misleading (optimistic) results in the **phenolphthalein** test

upgrading: *see* **improvement**

'Van Daveer' criteria: 4.5.5: empirical criteria relating the likelihood of corrosion to the surface **electrode potential**

vulnerable locations: *see* **checklist of vulnerable locations**

warranties: 5.8.2: *see* **guarantees**

waterproof membranes: 5.6.3; 5.12: faults in these can be locations for chloride intrusion

welding: 5.7.2

RECEIVED
JUL 8 2003